The Covid Letters

The Covid Letters

Alan Kennedy

Lasserrade Press

Lasserrade Press

www.lasserradepress.com

ISBN 978-1-9996941-6-6

Cover Image Anastasia Shuraeva

First published in the United Kingdom by Lasserrade Press 2024

For Elizabeth

Contents

Chapter 1

Blythe Beginnings - January to March 2020

These email letters began in early 2020, when a "scientific covid orthodoxy" scarcely existed. In any case, even if there was such a thing (and yes, I now think there was), it had no bearing on my own preoccupations at that time. In fact, I spent much of that year writing a psychoanalytic study of some of the novels of the children's author Arthur Ransome, a book greatly concerned with lying (a concept destined to loom large in these letters).

I am old enough to remember the epidemic of 'flu that struck Dundee in, I think, 1968. I recall my GP popping in, staring down at my fevered form and saying, "I think we'll call it 'flu – you will try and look after yourself, won't you?" (Those were the days). I have managed, since then, to contract whatever virus du jour was currently on offer. I think I've had them all, and the email dated 24 March 2020 suggests I caught this one, complete with its own made-in-China peculiarities. However, no doctor came visiting.

Searching for the Doliprane in the bathroom, I remember my wife cheerily reminding me of an exchange she'd had some months earlier outside the local *Maison de Santé*. (You have enough clues now that I live in France). The place was shut up, a notice tacked to the door saying its complement of medical staff was no more - they had (as the French say) displaced themselves. She asked the chap sweeping up outside what was going on and he said rumour had it they had all gone back home to Romania, but nobody seemed all that sure. All we knew was that we didn't have a doctor any more. Neither did anyone else in our small corner of rural France. We all have covid flashbulb moments – this was mine, the first indication that something very odd was going on. Just how odd, we shall see.

The letters are addressed to:

Professional psychologists (P)
Lawyers (solicitors, barristers, academics) (L)
Authors and artists (A)
(Relatively) normal people (O)
Mass mailings (for a brief period) sent to my "covid group" (GRP)
My wife, Elizabeth, sometimes appears as "E"
"Caesar" is a dog

I provide only my side of the exchanges, which have been lightly edited to remove all possibility of identification.

To L 6 Jan 2020.

A quiet Christmas was in prospect for us, so we went a little mad and decided to cook a goose (Julia Child's Vol 1 braisée recipe with chestnuts). I made the stuffing and retired the field, completely exhausted. Apparently, you can feed ten from a goose. We worked at it.
France is riddled with strikes and protests at the moment and one consequence has been that E's big present (special tree pruning gear), although ordered in November, has yet to arrive. Fortunately, there were reserve offerings, including special insect-proof clothing - very practical, my spouse.
Attached you will find your virtual card. Black and White and oven-ready for printing. It is [part of] a drawing by Arthur Ransome and [part of] a poem by Edward Thomas. All is explained in Chapter 3 of a book in progress.

To P 20 Jan 2020.

Thanks for sending the piece about the UK Health Service. Remember, I live in France – things are different here. I have a horrible suspicion it's somehow all too late, that one should start again - if that were at all possible. I fear the truth that will not speak its name is that people who count (you know - the ones who will get to drive the cars in the brave new world) no longer have to get their health care from the UK NHS. I'm sure there's private and private. (I've been reading a lot of Robin Collingwood and it's made me somewhat more cynical than usual).

Interesting you should extol the virtues of data. But there's something that must follow the data: viz people who know how to analyse them. As one of my mentors once said: don't expect the statistics to tell you what the mechanism is if you don't know already. He wasn't joking.

To A 2 Mar 2020.

Message and attached novel received.

I'm striving at the moment to complete an article for the Ransome Society (it's half of a chapter of the book) so I'll convert your manuscript to something that can inhabit my kindle. I'll read a little each night. It looks interesting. Give me a week.

You're right, though - the Ransome book is consuming my life. I've a deadline of 30 April. We'll see.

We're all signed up to an elaborate set of contracts with Orange and have joined the connected world, complete with "Le Livebox". It's plugged in but we shall rarely use it. E has her mobile working. I have simply determined to live my life out much as before.

I realise I've not heard from you for a while - it just goes to show what pounding keyboards does for you. I hope you are okay and that you avoid the dreaded virus. In a curious way all that stuff reminds me of the Great Ash Cloud that brought the world to an end a few years back. Fishy. Meanwhile, we are well prepared, with the larder full of bottles and tins. Enough for a week or two. A chap delivered some plasterboard the other day and I said I'd better not shake hands with you. He looked at me as if I was mad - Les Anglais.

To A 2 Mar 2020.

I'll have a look at your draft on Saturday, if that's okay. Can you send me instructions as to what bits I'm to read - I have not had internet for a while and see a lot of emails with attachments.

I've been placed in a category (age >70 and underlying heart condition) that means from today I am to be cut off from all human contact. I think it's for two weeks, but it is difficult to get sensible advice on the duration. All this is right in the middle of major building works going on right outside my study door. Sometimes life seems a real bugger.

You will probably be unaware of the mysterious recent failure of the Orange internet service for almost 24 hours. I must admit it was rather scary - not only no email and internet, but no banks, no post office and no local services.

I'm not allowed to go outside, but E tells me that the roads are very quiet. Did you see the re-make of the invasion of the body snatchers? Donald Sutherland in great form. I fret she will come back from the shops changed ...

I now intend to stand at the garden gate with my new smartphone and wave at the passing cars.

Virus: since it's a life form (is it?), I can't see why the ecoloons are not agitating against efforts to eradicate it. Not as pretty as a polar bear, I grant you, but all God's creatures and all that. Isn't biodiversity the name of their game? I take comfort from the fact that the Democrats are calling it Trump's virus.

To P 2 Mar 2020.

Since AK is > 70 (he flatters himself with that number) he can go nowhere at all and is confined to the house for an indefinite period. At first, it seemed a truly terrible restriction, but we now realise he never went anywhere, in any case. So, nothing much has changed. In fact, he can mow the lawn more often.

Actually we both intend to cheat a little and walk in the woods with our dog Caesar. They are our woods and we have never met anyone in them apart from strange men with guns hoping to shoot robins. Presumably, being small, they represent a challenge - it's hard to imagine there is much to eat in a robin.

I am going to re-read Mann, Camus and Pagnol on "La peste". Plus ca change.

Avoid people, breathe carefully and wash your hands.

To P 12 Mar 2020.

Thought you might like to see the form we have to fill in each time we leave the house on pain of paying a tidy "amende". Crikey... We're going for *aux besoins des animaux de compagnie*, but as you see, Caesar's well-being hangs on a comma.

The local Gendarme is a woman called Virginie. She's very nice and may even let us off. Our neighbour was a Colonel in the Gendarmerie (part of Chirac's *Garde du Corps*). Trouble is he lives in Bordeaux and probably feels he's stuck there! Otherwise, it's 138 euros a trip if we get stopped. I am looking into very complicated route that avoids the main road (unfortunately, me and many others).

It is quite extraordinarily quiet here. We saw a single car yesterday, slinking past and that was it. As for us, we go down the hill each afternoon, park the car, and wait while Caesar does whatever he does.

A friend who is secretary to the Mayor says we can walk in the woods there as a "couple", but we are all unsure how we get there (and whether we can get back!

We have started on the back catalogue of Medoc 2014. Excellent stuff - it will obviously go off if we just leave it.

Wash your hands.

To L 12 Mar 2020.

There have been very few cases here so far. Just a couple North of us, reported a week ago and a "suspected" in isolation in the county town, 100 km away. Of course, I may be looking the wrong way - we're not so far from the Basque country, but I can't understand (even a tiny bit) of their bloody language. All I know is it uses an unreasonable number of Xs.

We spent the afternoon in the garden learning how to send messages to each other on our new toys. I practised saying I shall be late getting home - one can but dream.

I hope to get back to Arthur Ransome tomorrow. So much drama these past couple of weeks I decided to catch up on research for something downstream. All done now and no further excuse available.

My regular Monday meetings to talk French have become virtual, using a CISCO Conferencing App. So far, plenty of takers. Actually, it suits me perfectly - I can hear everybody perfectly.

Weather very nice. 19 degrees at the moment with a lot of water creating a silvery glare in the air. It looks like no bedding plants this year. I've planted 300 tulips instead (E went a bit overboard buying). They are going to be everywhere. Sturdy little buggers - the slugs have given up on them.

One bright spot - everybody is burning like mad. It was made illegal a year ago and became completely clandestine, carried out at night. Now the place is like Apache country with rising smoke on the hills. We shall be joining in.

We've watched Bojo on the UK telly several times. Poor chap - hard to keep positive when he's being fed nothing but bad news (not that there's much else). A propos, I do wish the BBC would stop talking about exponential increases. Bloody Arts graduates. You'd think somebody would sit down and explain to them that you run out of people pretty fast with an exponential! An epidemic is usually described by a Gompertz function. I believe there was even a Mr Gompertz, although with a name like that he doesn't sound like somebody you'd want to know. Mind you, you get more information than we do. All we get is protracted periods of silence punctuated by stern injunctions with fines attached. I imagine the house arrest will be extended next week, probably for another month. Thereafter, it will be impossible to keep kids locked indoors. I shall put bait down in case they come this way. Gingerbread of course.

To P 23 Mar 2020.

Greetings from the plague house. We are in isolation - meaning we see literally nobody at all for days on end. Very rarely, Caesar gets very excited and rushes to the gate to bark at a car. Perhaps a car every two or three days. All this while the trees are bursting into leaf and flowers are poking up everywhere as usual.

France is adopting slightly different measures in response to the virus. You are not permitted to go out at all unless your trip falls into certain restricted categories, for which you need to fill in a form and carry it with you to show to the police. I imagine that is what life in Soviet Russia was like. We have filled in the form saying we are "walking the dog". We drive down the hill each day and let Caesar free so that he can run about for half an hour. He is puzzled that it is always the same place, but he'll have to put up with that.

Interesting to see that Saint Greta has had her prayers answered and the world has now stopped. "Capitalism"- as she dismissively calls it - is coming to an end. Whether she really anticipated the consequences is another question. It looks as if I may not be alive to know.

I have been reading a lot about the statistics of epidemics. They are generally fitted with something called a Gompertz function, discovered by an 18C chap called, oddly enough, Gompertz. It fits many biological events with initially slow growth, a period of quasi-exponential growth, and a rapid negative growth.

Oops - I am being called to deal with Caesar. He has a tick that has to be removed. More later ...

... That didn't take long. He's been lying on the lawn a lot in the sun and managed to pick up a tick. He is a very intelligent dog. Once he knows what you are doing, he lies very still and lets you carry out the operation. If you don't know about that, it involves a "tick hook" catching the body and rotating it so that the little beast is removed whole. Anyway, it's out.

Also alarmed you might be out of a job. If this is because of the virus, it could be you can claim salary. Goodhart's Law (look it up) suggests that there will soon be an extraordinary number of "virus-related" job losses in the UK.

I get news from my University now and then. They have cancelled examinations for the 1st and 2nd year. Perhaps they will never be restored - it would save complaints about grade inflation. I suggested years ago that all students were awarded a starred first-class degree on entrance and then had to spend four years justifying it by dealing with increasingly powerful challenges to their academic pretensions. A bit like being in a state of continual peer review. Who knows.

In about two weeks we run out of food and E will fill in a form and see whether she is allowed to visit a shop. I suppose they will let her. In any case we still have a lot of wine, some of it very good, and shall be converting to a liquid diet in the near future. Caesar has run out of carrots and it is too late to plant them. We are going to see whether he can eat cooked potatoes until the shop expedition.

Please say hello to husband. If he is looking for a new occupation, I think the icon business is due for a major up-turn. Saint Greta set in a sturdy wicker frame, tinted green.

A (not to mention tick-free Caesar).

To L 24 Mar 2020.

Glad it's worked out - these rushed things always seem to turn on the tiny unexpected events, like forgetting to fill up, losing a key or discovering there's only one match in the box. There's a very good Jack London short story about that.

I think Bojo's ambiguity was intended, even studied. Edouard Philippe (he's the PM – President Macron has vanished) made an almost equally infuriating speech yesterday lunchtime. It's clearly being done in concert. The bloke rattled on for half an hour about a necessary "hardening" of the rules until you felt like screaming. All anybody wanted to know was what *exactly* was going to change. And do you think he'd say? Not on your nelly - and we still don't know.

There is something in the newspaper this morning about not moving more than 1 km from an (undefined) reference point given the vague definition "chez vous" and a claim that you have to put the hour as well as the date on your Permission-to-Move Form. But the form to download is completely unchanged! Obviously, the concern related to selfish young people abusing their outdoor activity rights by enjoying themselves - at the seaside what's more! I really can't see the gendarmerie doing much to stop that here. There's not enough of them for one thing - and they are all deployed in the big cities not on the Atlantic coast. How the hell do you clear a five-mile-long beach full of drunks.

I assume Bojo's words were carefully crafted to be in lockstep with Paris and Berlin (just the answerphone in Rome) and to give a little window of opportunity for creative interpretation. I didn't stay up to watch him, although E did. We both enjoy watching a female journalist on Sky who is clearly in love with him, but I was too tired even for that. The little smile that flutters across her pretty little lips when she mentions his name … oh my.

Just back from Caesar's daily excursion. He absented himself for 40 minutes as punishment for our keeping him in yesterday while we removed a tick. Otherwise, lovely weather.

To L 24 Mar 2020.

I too was awake all night, having dined on razor blades. I dragged myself downstairs to slump in front of the telly. Watched Ferguson giving evidence to the Select Committee for two hours. His line now is that the value of R in the March 16 paper was, indeed, an over-estimate. And NHS capacity now very unlikely to be overwhelmed. I think I mentioned that seemed obvious from the start to anybody who could read. Perhaps he should have spent more time scouring the web.

It was interesting (at least to me) that fellow climate denier, Graham Stringer (who was not present) had sent in the only decent question, right at the end. It was about the sensitivity of the model to assumptions about R! "Good point" was the shameless reply.

Ferguson and the like represent a priesthood. They are dangerous people and are now fighting for political power. Politicians who wouldn't know a differential equation if it strangled them should not be lining up to ask "What does your model say about X?" As if the answer meant anything! Perhaps the one small candle in all this dark is that Bojo does appear to have had his doubts. I start to warm to him (albeit he's a bit of a lefty). He should have trusted his intuitions.

There is a paper by a psychologist called Lewandowski about conspiracy. He attempted to demonstrate that what he called "climate deniers" were prone to conspiratorial ideation caused by their indiscriminate web-use and endless drinking of fake-news (the customary smears, albeit it is easy to show that the exact inverse is the case - claims about fake news simply represent the old guard hanging on to their version of reality.) What became known as the "moon-landing" paper rested on counts of subjects endorsing (or otherwise) propositions about all the usual crank notions. It was widely cited (even by Obama). Trouble was, when the data were eventually released, they showed the reverse of what was claimed (of course). Not that he cared, he had made his point.

To P 25 Mar 2020.

I found the Ioannidis comment interesting (aka I agreed with it!). I wonder whether the stats quoted are correct. If you ever looked at the French analysis I sent you on 17 March, you'll see there were grounds even then for suspecting the chaps at Imperial played a blinder in terms of getting their own highly-controlling way. God help them when the post-hoc stuff is done.

Easy to predict where this is heading.

He/she is certainly correct to accuse Imperial (and LSE) of lefty scaremongering on "climate". As to their game on covid I am not in a position to know. Yet.

I had a very bad night, with what seems like the return of symptoms of my stomach ulcer. I've managed for a good while avoiding the bloody drugs - it looks as if I shall have to give in. I'll hold off and see what no-alcohol does. It's a delicate balance, because stress really does affect ulcers (notwithstanding the orthodox medical line - I have the papers). And these are stressful times for which wine is a blessed balm...

PS - "Is a Once in a Century Fiasco Befalling the United States?" - John Ioannidis, Professor of Medicine and Epidemiology at Stanford University, concerned that the nation is making dramatic decisions without reliable data. "The current coronavirus disease has been called a once-in-a-century pandemic. But it may also be a once-in-a-century fiasco.".

To O 25 Mar 2020.

Judith Curry is a serious academic. A former professor of climate science she gave evidence to the last US Senate Inquiry into the subject and influenced US policy. She is no crackpot and certainly believes in a "warming world".

Nic Lewis, albeit with no academic post, is a serious and respected statistician with a background in maths. He is taken seriously by the UK Met Office and (along with Curry) has produced a good (and thankfully low) estimate of "climate sensitivity". His current take on covid is that "UK modelling hugely overestimates the expected death rates from infection".

https://judithcurry.com/2020/03/25/covid-19-updated-data-implies-that-uk-modelling-hugely-overestimates-the-expected-death-rates-from-infection.

(Modesty forces me not to say Kennedy was on to this the day after)!

Conclusion: It is undeniable that the Imperial paper was seriously in error. The authors knew this - it was classic fake news. Professor Ferguson was challenged at the HoC Select Committee on the extreme sensitivity of his model to estimates (guesses) of values of Rzero (the rate at which one person infects others). Unfortunately, the people there didn't understand the question and the person asking it was not physically present! In other contexts, Ferguson's claim that without "lockdown" measures there would be millions of deaths would have been treated as crying "fire" in a packed cinema. He should be locked up. But Imperial are given to this sort of thing.

https://www.imperial.ac.uk/news/194711/
rising-temperatures-cause-over-2000-fatal.

Fortunately, along with the cruise ship, Sweden (where the response to the epidemic has been close to the original UK proposals), is going to provide a crucial control group. We shall see - I would like to be forced to eat my words but my libertarian bones tell me otherwise.

Sorry to sound angry but I have seen the literature on "climate suicides" among children. I even extracted a promise from the BBC "Complaints Unit" (that's a laugh) that they would bear this in mind. A subject for another day.

To P 28 Mar 2020.

I am starting to rebel.

I don't believe in a Climate Emergency. It doesn't deserve its capital letters. In fact, I think the idea is NONSENSE.

I don't believe in Universal Socialism. As an unreconstructed Methodist (a sect unknown in France) how could I? "Saint Paul was right" is my motto. Something unlikely to appeal to the Pope (even though this one is not Italian).

I believe in Brexit. After all, who does not want to be rid of the pesky Brits? Perhaps Frexit? Who would really want to join the Mafia.

I believe neither Mrs LePen nor Mr Macron (both appear to be extremely strange people). My cry is: "Can I have someone else?

I do not much believe in COVID19, or even COVID20.

So, you see, I have become a non-believer.

I have great faith in the French - please don't let me down. Surely there are already essays on the absurdity of the current wholly artificial plague? I have attached something E gave me to read. I am inclined to agree with the sensible (if emeritus) person. What would Camus say?

Now ... here is my current plan.

We are locked in at the moment. We hear the guards at night (their keys rattle) and have worked out their routine. Tonight, we intend to tunnel out (Caesar will help). God is on our side.

We have made civilian clothes, dyed with home-made dye extracted from a tin of SuperU plums. Quite convincing in the dark, if a little sticky.

I have made a beret out of a melon skin, dyed with black shoe polish. Very convincing although it has left a distinct line round my forehead.

I have forged two passes in case we are stopped. Cleverly stamped with an expired "Made in China" label peeled off the back of a radio. With luck (we all need luck) the border guard may not spot they are forgeries.

If we make it all the way to freedom, we shall send you a (coded) postcard. Please declaim it at the morning Appellplatz (2 metres, remember).

Liberté.

To P 30 Mar 2020.

Hang in there. And feel free to scratch - assuming it's not forbidden. Here, you need to fill in a form to do things.

Suddenly I realise how different are our legal systems. I'm really not much of a rebel but I was brought up a free man. Well, a free little boy and few really grow much past that stage. You can do anything, they told me, unless the law forbids it - so be a good boy. And I was, more or less. But now .... you can do nothing, they tell me (nothing at all!) unless the law permits it. So be a good boy and stand still. It's easy to be scared of that.

Am working like a demon on my Ransome book. I burrow down into another world and end up feeling a little (a very little) better.

To L 30 Mar 2020.

I enjoyed that. Brought back memories of reading *Rupert the Bear* in the coal cellar with my little chums. Happy days.

Have not seen a person for many days. Great improvement.

I read the modelling study that so irritated Mr Gove in bed this morning (there's a syntactic scope issue there, sorry). Very clever. I love Oxford Bayesian clever clogs. Somewhere buried in it there's a range on a variable of 20-80 percent. Makes you wonder, don't it? Why not go the whole hog? thought I. 0-100 and you won't be even a little bit wrong, not ever.

Don't forget to wash your credit card along with your hands. There is no obligation to sing but you never know who has been squeezed between those rollers before you. They say you can't use Vodka because there's not enough alcohol in it. "They" obviously don't know about the Polish stuff.

Now ... here's something to really worry you (or really to worry you, if you're into that sort of thing). Do you recall the plot of 2001? Not the book, the film ... The bit where Leonard Rossiter strolls out of *Rising Damp* into a spaceship. Complete with Russian accent.

To P 30 Mar 2020.

I think it is already impossible to get data that are comparable across countries. Has the UK designated it as a "notifiable disease"? which may distort the mortality data. I don't think anyone now believes covid is 3 times more contagious (than 'flu). The man at Imperial College who produced the scary prediction that changed UK policy used a figure of 2.75, predicting 500,000 deaths. His recently revised model used 2.25 and almost 6000 deaths - obviously something is very non-linear! these are absolute numbers (one person will infect 2.25 others, other things being equal). I have no idea what the equivalent is for "normal winter flu" but I think it is also quite contagious. This disease appears to be contagious, but not particularly lethal unless old and sick.

There is a paper much discussed in the UK at the moment by some zoologists in Oxford. I have read it. It is very clever, using Bayesian analyses to find a situation in the past which matches the best estimate of the current data. On that basis (and quite remarkably) they argue that a huge percentage of the population in the UK must already have been exposed, presumably without symptoms.

It is the fact that it is a very "political" disease that makes me wary. In the UK it is all about the National Health Service - a kind of Socialist state within a state. The claim that the NHS is the best way of organising a "free" health service is very questionable. We find the French service much better.

Anyway, I don't like being placed under house arrest for the sake of the precautionary principle. Start down that road and god knows where you end up.

To A 31 Mar 2020.

I should have said the gleeful piece was by a bird from the *Financial Times*. Read with tongs. What the French call "infox" (look it up).

Sweden still suggests (may soon confirm?) a serious overreaction. People who want to change "the way we all live" are usually wrong. Correction ... invariably wrong. They do, however, and in the main, tend to have beards.

French rules are being gracefully retracted by a second decree - the addition of three new categories of exception and a longer form.

We're off down the tunnel. The dog is keeping the guards distracted singing "keep the home fires burning." His solo.

To P 31 Mar 2020.

The Imperial people have produced a new paper about the response to covid-19. Among other things it addresses (pre-empts one might think) the Sweden question. I attach a copy. It is technical, but easy to follow. One might consider milking the "without me the show would have flopped" line a trifle speculative (even premature), but then show business is show business.

I don't know whether Nic Lewis has yet published a comment on this new modelling. I'll look. It will be useful because he is extremely good at it. When he does (if he does) I will pass it on. You recall he was critical of the fact that Ferguson didn't publish the code of the earlier model (still hasn't, as far as I know). This group does provide the code. None of this stuff has been peer-reviewed, of course.

I don't recall whether I passed on the update to Lewis' 25/03 paper (two days later), noting that the number of cruise ship deaths had increased by 2 (it did not affect the results). However, there is a footnote (number 12) which now seems quite important: he remarks that he cannot reproduce the 500k+ figure widely quoted from the initial Ferguson paper. We shall see.

All this reminds me so much of the endless discussions in the Royal Society of Edinburgh about the response in Scotland to the Foot and Mouth pandemic. Bill Stewart trying his best to stop mass slaughter. The report was unusually critical of the use of statistical models run by people with an agenda to drive policy. Nothing much has changed.

Here's a bit, spliced from the Appendix - Paras 85 – 92 Royal Society of Edinburgh Inquiry into Foot and Mouth Disease in Scotland, July 200.

"The Use of Models.

85. Quantitative techniques, including risk assessment, analytical epidemiology and modelling, are dependent upon good quality data that is accurate and timely. This includes accurate and timely diagnosis (see paragraphs 53-56).

86. Models can be used to describe past and current epidemics and to explore the course of ongoing and future epidemics. In view of the data that are available and the complexity of the epidemiology of infectious disease, any model will, at best, only approximate to reality. Whilst models can be useful, they should not be the sole dictator of policy nor should they remain unchallenged and unvalidated. Given that most epidemics are unique in one or more aspects, historical models might provide a useful framework but will inevitably require updating and the addition of local information relating to climate and topography will be necessary.

87. Plume models developed by the UK Meteorological Office and the Danish Meteorological office, used early in the outbreak, showed that cattle and sheep on a nearby farm to the pig unit in Northumberland were most likely infected from that pig unit and that the risk of long distance

spread of the disease to Europe through airborne virus was extremely low. Airborne spread over 3km was not found to play a significant role in the epidemic (21). Most spread within 3km ('local spread') was attributed to local aerosol spread between animals or contamination in the areas near the outbreaks. The model indicated that spread from cattle and sheep was minimal. However, this was criticised in evidence received, since it did not take account of the enormous quantity of foot and mouth disease virus in lesions, milk and faeces from cattle (22). Incorporation of such a source of the virus from cattle in a short range model would have helped to identify farms down wind at risk as well as local contamination.

88. Much of the controversy during the 2001 FMD epidemic related to the use of models to predict the course of the outbreak on which the intervention strategy was based. At the time of the outbreak, there was much criticism of some of the assumptions underlying these models. Some of those responsible for the modelling work argued that, without the action that followed, the epidemic would have been a great deal worse. Others, who submitted papers to us, took the view that the epidemic was already passed its peak before the contiguous culls that were based on the models took effect (23), albeit that modelling can assist in the process. Subsequently a generation of more complex and mature models have emerged which have benefited from more complete data sets and greater investment of time. However, they remain models and the scenarios they explore are only possible examples of what might have happened (24). A thorough analysis of the complete data from the epidemic must therefore be a priority.

89. In the context of quantitative epidemiology, modelling analyses are ideally performed in consultation by a team of people with the relevant quantitative, clinical, virological, immunological skills. For disease controls and policy to be effective there must be consensus, consistency and compliance. Stakeholder inclusion is critical to this success. Several of those submitting evidence were strongly of the view that there was neither consensus among informed experts nor stakeholder inclusion in 2001 (25).

90. It is challenging to model an epidemic in which there is a large amount of spatial and temporal variation, either due to regional differences or changes brought about by the interventions occurring during control. It is particularly demanding to develop a model which explores the impact of a range of possible interventions where complex biological systems are necessarily represented by a few variables. It is highly likely that unmeasured or unknown variables and factors exist. Transparency is therefore of the utmost importance. In the 2001 epidemic

one of the most influential models was not available for inspection by the wider scientific community until it had been published in the scientific press after it had informed policy (26).

91. It was not explained how scientific advice was solicited in either the UK or Scottish context, despite recommendations existing for such procedures (27). There was a large body of epidemiological and modelling expertise in Scotland, in research institutes and the universities, that was not called upon during the epidemic. Similarly, data management skills and land use databases existed that were not immediately accessible. We received evidence that geocoded data regarding animal distribution were flawed (28).

92. In the epidemic of 2001, some reassurance can be taken from the fact that the several different models largely came to the same conclusions. But as very few of the contiguous premises were tested it is impossible to tell whether this aspect of the policy was justified."

Chapter 2

Darkening skies - March to May 2020

By the spring of 2020 I had drifted into one of the very many different groups trying to come to terms with the outbreak of "covid". We found ourselves confronted by claims that were, in turn, baffling, contradictory, nonsensical and (ultimately) sinister. These email letters define my membership of one particular group: quite difficult to characterise, I would describe it as "scientifically literate, well-meaning, interested outsiders with no particular axe to grind." A definition that with hindsight seems dreadfully naive, but I was not to know that. As to "scientifically literate," I may make that claim. Psychologists come in many different flavours, spanning a range from, on one flank, "almost-a-sociologist" to "almost-a-neuroscientist" on the other. I am closer to the latter than the former, having been deeply concerned with computational models of eye movement control for many years.

Viewed from France – a country which had experienced a number of recent health-care scandals – the initial UK response to this allegedly novel respiratory disease seemed like an overreaction. Reports of an outbreak of the disease in mid-March 2020 on board the French aircraft carrier *Charles de Gaulle* appeared to confirm this: around 1700 of the crew – fit young men in the main - contracted the disease, but there was little call on the ship's hospital facilities, little serious sickness, and certainly no fatalities. Approximately a fifth of the passengers and crew on board the cruise ship *The Diamond Princess* became infected and, in this case, at least 14 were reported to have subsequently died. The fatalities were however, old and in poor health and died *with*, although not necessarily *from*, the virus.

The letter written by Dr. Sucharit Bhakdi (6 April) echoes my own feelings of growing alarm over the interpretation offered by politicians of wholly unexceptional data. Even given Dr Fauci's anodyne reassurances (see the letter dated April 15th), unprecedented actions were set in motion, including house arrest for much of the population of France. It was surely necessary to ask *why?* and difficult to avoid the conclusion that the extraordinary response was justified, not by any particular disease, but by something less tangible, almost mythical – the unfolding of an apocalyptic catastrophe born of the fevered imagination of Hollywood movies. In the course of a few weeks, "covid" became as tribal and as fatal an intellectual issue as "climate".

To L 31 Mar 2020.

It is in Para 88, but that whole section is correct. Nobody did anything of course.

I understand the Imperial model was based on 13-year-old code developed for the 'flu outbreak of that year (warning: this may be false, but the source looks reliable).

Over and out.

To A 1 Apr 2020.

You do seem to go out far more than we do. In fact, we've only had one expedition since the 16th. E to SuperU. She found it much as usual, although she was so stressed she forgot to buy several things. She's aiming to drive to the local town on Thursday, so may well be arrested. Her new cheque book is lying in a box in the Post Office there and she really needs it. We've been trying to fit her excursion into the newly extended list of "things you're allowed to do ...". I print one out each day, fill it in, and drive Caesar down to hill for his walk. No Gendarmes so far.

God alone knows what to make of the UK response to all this. I suppose there will be the usual post-mortem Report one fine day. It does look like an overreaction. On the other hand, I gather it's a disease that likes ancient people with dicky hearts, so maybe I'll not get to read about it. The mismatch between what the statistics seem to be saying and the general air of political panic is itself quite scary. I recall watching nice Mr Blair going on about Weapons of Mass Destruction years ago and saying to E, "he must know something really dreadful." He didn't, of course. He was just lying. But you can't help that same "what do they know?" feeling, can you?

A relative's wife arrived in the middle of the night and was deposited in some far-flung bit of their empire where the gardener usually makes his tea (he insists she "isolates"). She then managed to lock herself out taking a trip in the dark to the loo. At 3 am. Windows had to be broken, everyone standing 2 metres apart (they could have stood closer in France). It sounded like something the Whitehall theatre used to do. I sent them my favourite spoof Peppa Pig episode to cheer them up (Peppa Pig gets Corona Virus). It didn't.

Who's Mrs Hislop? Me, I'm downloading free books from Project Guttenberg onto my Kindle as fast as I can. Lots of early Wodehouse are now out of copyright. I've discovered a Psmith I'd not read. Not very good, but good enough in parts to take my mind off things.

Best to both of you. Be brave. Wash your hands. Pray.

To L April 3 2020.

A reader's comment in today's Le Figaro. I think the numbers come from the usual "World in Data" site.

> "L'Allemagne (83 200 000 d'habitants / 77 558 cas / 891 décès) parvient à limiter le nombre de décès. Pourquoi? Rappel : France: 67 064 000 habitants / 56 989 cas / 4032 décès!".

He has a point. France and Germany are roughly the same size and roughly speaking their medical systems are equally good. Both have cases roughly 0.1 percent of population. There are FIVE times the (reported) deaths in France. You don't even need to do the t-test! What on earth is going on? I only ask.

And no, it can't be more tests nor more respirators. Agreed, there are fewer tests (=cases?) in France but nothing like by a factor of five.

And there are plenty of ventilators in France according to the PM last night, although some are not always in the right places. But patients can be being moved about in quite a slick operation that's going well (they are even sending the buggers here!). So it can't be that, either. I saw a piece in *The Spectator* at the weekend by a medic raising the same question. Eventually, journalists will start asking.

Remarkable interview on UK Radio Four this morning in which the ex-CMO for Scotland let a cat out of the bag, saying nobody (aka him) had the faintest idea of the baseline infection rate in the population. You'd think somebody by now would know how infectious this disease is (Professor Ferguson claimed to know it to the first decimal place.) He went on to imply that the only way to know this was by testing "everybody". He must have skipped the session on sampling at his Medical School all those years ago. 10,000 should be enough - most clinical trials have less than that. Even I could do it for him. Why does he want it to be so hard? One thing I can tell him already - it is vastly more lethal in France than in Germany. *Why*?

E off to shops this morning, complete with mask. More shops opening. Even the Garden Centre. She returned with herbs. Tractors back in the fields, neighbour moving his sheep about (even waving from a distance – neighbour, not the sheep). The odd car rumbles by. It's very quiet here in any case, but starts to feel normal.

Politics - I only read *Sud Ouest* (left wing) and *Le Figaro* (centre right of the "bring back Chirac" variety). Going by them (and it's a reasonable sample). Mr Macron may be a one-term President. That'll be two in a row! The frogs do communal angry very well. Although "the left" initially seemed strangely vindicated by this disease - as if we all deserved it for liking Mr Trump - they have not, as yet, found a point of political leverage. Probably because it's not there. I keep penning missives to people pointing out they may (irrationally) dislike Mr Trump because he makes rude jokes about wimmin, but he has the best medical advice in the world. The US has more Nobel Prizes in medicine than hot dinners. (That's why he's right about "climate", incidentally.

Lots here about the scandal of the masks. Billions on order from China although a Chinese-owned factory in France is still exporting them. To China! Not that anybody is willing to claim masks do anything to alter the course of an epidemic. There has been a lot of research on the subject and the answer seems negative outside a hospital (where gowns etc are a confounding factor). They do look scary though and will provide useful cover for those who will need it when the stats are in. And that's what counts. Saint Greta will shortly be up from her sick bed sporting the eco-mask her dad is right now designing.

E (her price is beyond rubies) also laid down a mighty stock of *epinard* some time ago. She has a trick way of squeezing the water out then cutting it up with pizza scissors - very cute. Popped ready-buttered in meal-sized packs in the deep freeze. Some will be consumed this very day with leg of lamb.

We struggle on. Although he himmed and harred about it, M Philippe implied he'll let us out in May. If it's a hot summer, he'll have little choice.

Sweden still looking too much like Norway to be very consoling to the change-your-way-of-life brigade.

Here's something for you – go for it.

RAPID ASSISTANCE IN MODELLING THE PANDEMIC: RAM.

A call for assistance, addressed to the scientific modelling community and coordinated by the Royal Society.

[Editorial Note: This project generated a frenzy of activity but largely petered out in August 2020, with many forlorn web links left broken. One significant outcome was a Special Issue of the periodical, *Epidemics* (Volume 40), published in September 2022. Papers on Non Pharmaceutical Interventions, but (strangely) very little input from, or reference to, Ferguson or the people at Imperial. However, to quote one of the authors, "The silver lining to this dark cloud is that it has forced us to think more proactively about the whole picture of infectious diseases, and maybe even about the related existential threats of climate change and environmental destruction."]

To A 3 Apr 2020.

The Royal Society has put out an urgent appeal for modellers in ANY domain. Project to be coordinated from Cambridge (quite a significant scientific/political decision).

Arthur Ransome is easier.

To P 3 Apr 2020.

You answer a question I was not asking. I am not doubting that there is an epidemic in France. It seems not (yet) very different from a particularly bad seasonal 'flu epidemic (in which I gather about 25,000 may die) but it may turn out much worse than that. Nonetheless, my question stands. Why is the disease causing more deaths in France than in Germany? It was the question buried in Professor Lee's piece and it puzzled me. It is a very striking difference and it is surely the same disease. I can't think of a reason. His reason, as you saw, is that the data are not comparable, but he doesn't expand on the point.

The answer I get from other people I have asked (including from a senior UK Health Service Administrator) is that "somehow" Germany is better at dealing with disease or that Germans are possibly more compliant with regard to social control. I don't believe the first. There are data on the second in the form of car-use and it shows little difference. France has been generally very compliant. I conclude the data are not comparable across countries. That is extremely bad news.

Tomorrow I shall stop thinking about this question and return to my book about Arthur Ransome.

My beloved spouse went shopping today, complete with her mask. We have been locked in since March 17. I waited for her return pacing up and down, wondering whether she would be arrested for not completing the form correctly! What a life.

To "GRP" 5 Apr 2020.

Rather a lot of reading today ...
https://swprs.org/a-swiss-doctor-on-covid-19.
   Obviously all this is from someone who appears to think that the covid story is an overreaction. I am drawn to this position, but aware that I am (a) "fragile" and (b) unlikely ever to qualify for an Immunity Certificate. So more than a little frightened.
   These are not reasons to ignore the above completely, because some of the links are very informative - for example, the remarkable observation that the specificity of covid (attacking the old, sparing children, attacking those with underlying conditions) matches the "normal" mortality figures for all countries (except possibly North Italy). I wonder whether this is the canary in the mine.
   There are useful comparisons between countries here, in particular some commentary on the French data (they begin to look like outliers. Why?). Quite simple analysis of the famous "World in Data" website shows that the data collected in different countries are plainly not strictly comparable. This is extremely bad news.
   A second motif in this string of links and references is the obvious point that what is being tracked is "tests" as much as "cases". I have spent several hours as an amateur statistician trying to understand why the death rate in France should be so much higher than in Germany.
   The third motif is the "of versus with" debate. Perhaps the most critical issue of all. If a person who dies of a broken neck can nonetheless, be placed in the "with" category, we ended this business with datasets that are close to useless.
   Finally, as a veteran of the climate wars I find the increasingly "political" nature of this epidemic/pandemic unnervingly familiar. (In my opinion, the claim that god votes Labour is a little like the claim she speaks English).
Anyway, if you are locked in on a sunny afternoon this stuff might while away the idle lockdown hours. I attach a version, but you will need to access the website itself to access the links (some, although not all, of which I found illuminating).
   I need to say that we climate veterans are inured to the charge of "conspiracy theorist". My response is two-fold: First, Dixon and Jones

(you'll have to look it up). Second, some years ago I found myself in a sort of academic "war of models." It was a little depressing but left me more than usually wedded to experiment.

I am scaling back the bulletins a little, to get on with Mr Ransome.

To A 5 Apr 2020.

Thanks. Yes, I'd seen it. There is also a lot of stuff about corruption in the WHO. Certainly, the daily data deluge has become a complete farce, with no context whatsoever. It all seems something of a swamp-by-design, does it not? I seem to recall a similar racket involving Tamiflu and the WHO. I got something from my university a little while back. I'll try and find it. It shows how horribly dependent we are now on Chinese postgraduate students and on the new "Life Science" priesthood. Serves us right, I suppose.

Mind you, if Gates' vaccines are as good as his software, we can sleep easy.

To "GRP" 6 April 2020.

An Open Letter from Dr. Sucharit Bhakdi, Professor Emeritus of Medical Microbiology at the Johannes Gutenberg University Mainz, to the German Chancellor Dr. Angela Merkel. (I circulate because he asked.)

Professor Bhakdi calls for an urgent reassessment of the response to Covid-19 and asks the Chancellor five crucial questions. The letter is dated March 26. This is an unofficial translation; see the original letter in German as a PDF.

I'll just paste it here.

"Dear Chancellor.

As Emeritus of the Johannes-Gutenberg-University in Mainz and longtime director of the Institute for Medical Microbiology, I feel obliged to critically question the far-reaching restrictions on public life that we are currently taking on ourselves in order to reduce the spread of the COVID-19 virus.

It is expressly not my intention to play down the dangers of the virus or to spread a political message. However, I feel it is my duty to make a scientific contribution to putting the current data and facts into perspective and, in addition, to ask questions that are in danger of being lost in the heated debate.

The reason for my concern lies above all in the truly unforeseeable socio-economic consequences of the drastic containment measures which are currently being applied in large parts of Europe and which are also already being practised on a large scale in Germany.

My wish is to discuss critically - and with the necessary foresight - the advantages and disadvantages of restricting public life and the resulting long-term effects.

To this end, I am confronted with five questions which have not been answered sufficiently so far, but which are indispensable for a balanced analysis.

I would like to ask you to comment quickly and, at the same time, appeal to the Federal Government to develop strategies that effectively protect risk groups without restricting public life across the board and sow the seeds for an even more intensive polarization of society than is already taking place.

With the utmost respect.
Prof. em. Dr. med. Sucharit Bhakdi.

Infection

In infectiology - founded by Robert Koch himself - a traditional distinction is made between infection and disease. An illness requires a clinical manifestation. Therefore, only patients with symptoms such as fever or cough should be included in the statistics as new cases. In other words, a new infection - as measured by the COVID-19 test - does not necessarily mean that we are dealing with a newly ill patient who needs a hospital bed. However, it is currently assumed that five percent of all infected people become seriously ill and require ventilation. Projections based on this estimate suggest that the healthcare system could be overburdened.

My question: Did the projections make a distinction between symptom-free infected people and actual, sick patients - i.e. people who develop symptoms.

2. Dangerousness

A number of coronaviruses have been circulating for a long time - largely unnoticed by the media. If it should turn out that the COVID-19 virus should not be ascribed a significantly higher risk potential than the already circulating corona viruses, all countermeasures would obviously become unnecessary. The internationally recognized International Journal of Antimicrobial Agents will soon publish a paper that addresses exactly this question. Preliminary results of the study can already be seen today and lead to the conclusion that the new virus is NOT different from traditional corona viruses in terms of dangerousness. The authors express this in the title of their paper SARS-CoV-2: Fear versus Data.

My question: How does the current workload of intensive care units with patients with diagnosed COVID-19 compare to other coronavirus infections, and to what extent will this data be taken into account in further decision-making by the federal government? In addition: Has the above study been taken into account in the planning so far? Here too, of course, diagnosed means that the virus plays a decisive role in the patient's state of illness, and not that previous illnesses play a greater role.

3. Dissemination

According to a report in the Süddeutsche Zeitung, not even the much-cited Robert Koch Institute knows exactly how much is tested for COVID-19. It is a fact, however, that a rapid increase in the number of cases has recently been observed in Germany as the volume of tests increases.

It is therefore reasonable to suspect that the virus has already spread unnoticed in the healthy population. This would have two consequences: firstly, it would mean that the official death rate - on 26 March 2020, for example, there were 206 deaths from around 37,300 infections, or 0.55 percent - is too high; and secondly, it would mean that it would hardly be possible to prevent the virus from spreading in the healthy population.

My question: Has there already been a random sample of the healthy general population to validate the real spread of the virus, or is this planned in the near future?

4. Mortality

The fear of a rise in the death rate in Germany (currently 0.55 percent) is currently the subject of particularly intense media attention. Many people are worried that it could shoot up like in Italy (10 percent) and Spain (7 percent) if action is not taken in time.

At the same time, the mistake is being made worldwide to report virus-related deaths as soon as it is established that the virus was present at the time of death - regardless of other factors. This violates a basic principle of infectiology: only when it is certain that an agent has played a significant role in the disease or death may a diagnosis be made. The Association of the Scientific Medical Societies of Germany expressly writes in its guidelines: In addition to the cause of death, a causal chain must be stated, with the corresponding underlying disease in third place on the death certificate. Occasionally, four-linked causal chains must also be stated.

At present there is no official information on whether, at least in retrospect, more critical analyses of medical records have been undertaken to determine how many deaths were actually caused by the virus. My question: Has Germany simply followed this trend of a COVID-19 general suspicion? And: is it intended to continue this categorisation uncritically as in other countries? How, then, is a distinction to be made between genuine corona-related deaths and accidental virus presence at the time of death.

5. Comparability

The appalling situation in Italy is repeatedly used as a reference scenario. However, the true role of the virus in that country is completely unclear for many reasons - not only because points 3 and 4 above also apply here, but also because exceptional external factors exist which make these regions particularly vulnerable. One of these factors is the increased air pollution in the north of Italy. According to WHO estimates, this situation, even without the virus, led to over 8,000 additional deaths per year in 2006 in the 13 largest cities in Italy alone. The situation has not changed significantly since then. Finally, it has also been shown that air pollution greatly increases the risk of viral lung diseases in very young and elderly people. Moreover, 27.4 percent of the particularly vulnerable population in this country live with young people, and in Spain as many as 33.5 percent. In Germany, the figure is only seven percent. In addition, according to Prof. Dr. Reinhard Busse, head of the Department of Management in Health Care at the TU Berlin, Germany is significantly better equipped than Italy in terms of intensive care units - by a factor of about 2.5.

My question: What efforts are being made to make the population aware of these elementary differences and to make people understand that scenarios like those in Italy or Spain are not realistic here?

To A 7 Apr 2020.

Spent all day working at something less slippery. I like Boris and he seems quite well-educated - at least compared with the competition. I'm sorry he's in such great peril. I hope he pulls through. I once spent ten days in an ICU containing six bays and during my sojourn five residents died. Some noisily. I don't think I shall recover from the experience.

Trump will come out of this quite well, I think. He has very good advice. The best, in fact. All he has to do is listen. You can't really see the US going for Sleepy Joe. At least I can't.

My university is pretty well owned by big pharma (and there are some spectacular people including a Nobel). But it's been at a price. As the years have passed all the other disciplines have withered. And those left are all so woke - they cling to social justice like a surrogate mother. It grieves me. Like most small institutions we depend on overseas PG income, mostly Chinese nowadays.

I conclude Covid has been a hoax of some kind. A bit like "Climate". Apply the precautionary principle thoughtlessly to anything at all and you end up in queer street. I see the BBC resident climate loony is now suggesting all the planes are left grounded. How I'd like to switch his phone off.

Help.

To P 9 Apr 2020.

A couple of things to look at in these dark days.

First an interesting site set up by an architect / amateur statistician, trying to make sense of numbers. Therefore, a man after my own heart. He has made a few mistakes (he has that in common with both me and Professor Ferguson) and his graphs are sometimes naive. But the data he uses come from the ONS and must, therefore, be accurate :)). I find them quite startling.

The second is a piece by Toby Young in *The Critic*. Yes, that Toby Young - but well worth reading. It's on the economics of disease.

https://thecritic.co.uk/has-the-government-over-reacted-to-the-coronaviru.

To P 10 Apr 2020.

I've been a bit swamped book writing all day. Fun in a way, but another world.

There is an enterprise (very kosher and not conspiratorial at all) called something like Oxford Centre for Evidence Based Medicine. Look at their data on the case mortality rate (cases/death) for different countries. The range of values is quite MAD. Trump is equally mad, but (alarmingly) right about certain things.

To A 13 Apr 2020.

He's no friend of mine [Boris Johnson]. Too much into climate.

I see Kissinger has risen from wherever he had been stored to say something about the New World Order. Old men and their passions.

To A 15 Apr 2020.

Interesting (quite recent) article by Dr Fauci. Yes, that Dr Fauci. As you know, the Euro Morbidity figures continue, in fact, to support the notion that this disease can reasonably be compared to "seasonal 'flu." Perhaps the response seems an overreaction. If so, I told you so. I think the critical bit is Para 3 or 4.
https://www.nejm.org/doi/full/10.1056/NEJMe200238.

Judith Curry has a (too long?) article on what she calls "epistemic trespassing." bringing together (inevitably, since that's what this is all about) "climate" and "covid".
https://judithcurry.com/2020/04/14/in-favor-of-epistemic-trespassing/.

If you read it I can't help pointing out that yours truly, who really does know the hell of a lot about "climate", is usually dismissed as a crank trespasser-psychologist, off his turf, if not his rocker. Whereas Sir Paul Nurse, the molecular geneticist has helped determine UK "Climate" policy for years ... Ah well.

Finally, if you have the stomach for it (he rides well to the right of centre, where it is dark at noon), take a look at William Briggs (Statistician to the Stars). He is, alas, a Bayesian, but awfully clever. He has a whole series of excoriating blog postings, well worth the reading. You will have to search.
https://wmbriggs.com.

We inhabitants of France (excluding the political class and the gendarmerie) are now under continuing house arrest in an on-going test of what future generations (should there be any) will know as Greta's Conjecture.

To L 15 Apr 2020.

Okay, sorry to cause offence. I'll drop you off the distribution list. I still suggest you read the Fauci article. He isn't a crank (I don't think he loves Mr Trump, if that's a recommendation).

Deaths can be counted. Not dying really can't be counted sensibly. And "averted" is simply a tendentious concept. It is statistically, impossible to know (you'd not get ethical approval for the necessary experiments!) But one thing's certain: it can be a HUGE number given a big enough initial prediction.

In terms of outcomes the outcomes look very similar to 'flu. All we have is the numbers (epistemic trespassing does permit that: even I understand numbers). Look at the EUROMOMO site. European mortality by age, by year. It's not even elevated in the older age group. I suppose you could say "yet," but most European countries say the peak is passed. The figures suggest UK not exceptional. The worst year of deaths from seasonal flu 2016 (week 48) - 2017 (week 8). We are still well below that in all age groups (and currently going down, although it is clearly elevated). I gather it's like that pretty well every year. New York is the same. Underfunded Health Service and lots of poor people crammed together.

But the Fauci article suggests the "case fatality rate" of this particular disease is the same as "seasonal 'flu". Not my opinion - that's what he says. It's no good reason to lock me up.

Sincerely yours, (from house arrest).

To P 16 Apr 2020.

I will happily do this. It will be a break from writing about Arthur Ransome.

A word of caution, though. I am in the sceptical camp about the utility of social distancing and locking people up. The state of the data is frankly a scandal. As is the fact that quite simple and obvious measures (e.g. random sampling for evidence of prior infection / immunity rates) have not been carried out (or at least not reported). The Government is complicit in allowing the use of very misleading graphs, with no context at all. Is 2000 deaths a lot? More than usual? Less than usual? Where are the comparative data that allow you to draw sensible conclusions. If 60,000 people died of 'flu in 2017, would that alter people's opinion of the severity of this epidemic. Why is this disease not declared officially as "high risk"?

Virtually everything about this disease is a muddle. The govt is behaving as if it's an outbreak of MERS, when the evidence (thank god) suggests its "case fatality rate" is much the same as normal seasonal 'flu. Which is correct? There are often bad outbreaks of 'flu in the UK (2016/17 for example) with greater death rates, yet previously there has been little or no newspaper coverage and certainly no draconian control measures that risk breaking society and the economy altogether.

As I understand it, the models used to forecast mortality rates, like the famous model of Professor Ferguson, include a great deal of amateur social psychology. By which I mean gross assumptions about "behaviour". Should experimental psychologists not have been asked to consider this? Even to have some input into the process? The discipline of social psychology itself has not (so far as I know) adopted the current fashion of modelling everything. Is it not worth asking why? I assume (again, I may be wrong) this it because social psychologists realise it's just too complicated. Social behaviour may not be deterministic in that sense at all.

Just so you know what I believe! We appear to be in the middle of a pandemic of mass hysteria.

To L 16 Apr 2020.

As I said, we disagree about averted deaths (aka "lives saved"). If excess deaths are a proportion of all deaths (which can certainly be counted) there is no issue. That is, they are simply a subset of the set [all deaths]. If they are not estimates (i.e not guesses about some future unknown state) they can certainly be counted. Now, if by "estimate" in this context you simply mean the assigning of members of the set [all deaths] to some category defined post hoc as [excess deaths] then we are simply using the word "estimate" in different ways. Estimates of "excess" deaths are now reduced to claims like "of all the deaths which occurred I deem these to have been unnecessary." The argument is reduced to debate about the category, NOT the count.

The claim "you are alive, but had you not said Hail Mary thrice you would be dead now," is what I term a claim about an averted death. The table containing these may have columns and rows and be as big as you like to make it - and some like it really big. How about 500,000 for starters? If by estimates you refer to its more usual sense as relating to some predicted/projected state (e.g. I estimate your chance of being alive tomorrow as 0.999) such probabilities are not "events". They can very definitely not be counted. Not if you mean by "count" what "count" means. Of course, by tomorrow you will be either dead or not dead - there's no other alternative – and can be counted. Remember the story about the little boy who sounded the horn to keep the bears away? Do we really want to be talking about "counting" his averted bears.

To P 18 Apr 2020.

In my spare time (ha!) I've been reading the famous Imperial paper. Attached. The model itself is in Figure 5. I'm not clear how the interventions were shared out across countries or how (if) they were weighted, but I assume it was reasonable.

I've got some specific questions. They no doubt reflect my lack of understanding.

The model is written in R and is on the web. The data set is surprisingly small.

Can I test my worries out on you? They all relate to validation (in different ways).

First Figure 8. Does what they say about this figure really match the data presented? The log scales for X and Y are different.

Second. The test with an uninformative prior Figure 8.4.3. If one intervention is acting as a proxy for others, isn't this a serious difficulty? I see there's a (throwaway) line about collinearity, but this seems like a definition of it! I am puzzled. Going back to the model itself there is no way of seeing how sensitive it might be to this fact. I do see, of course, that if you lock everybody up some of the other interventions become pointless, but they evolved in time. It's the effect of this in the model I'm interested in.

Finally Figure 18. I don't understand what they are doing here, so would appreciate an explanation if you understand. It is badly written but seems to imply that their values for the initial value of R can be validated by comparing them with the output of a "simpler model". I don't follow the logic, but that's my problem. I do know, however, from reading other things, that this initial value of R pretty well determines everything else. So, my issue is the claim that the graph shows a "large correspondence" between the two estimates (red and black). It obviously does no such thing. So, am I misreading it? Are they talking about good correspondence with some other value of R presented earlier in the paper.

You may not want to think about all this. I am only interested because it so much resembles the arguments about foot and mouth disease modelling in Scotland many years ago. I was involved in that and kept some of the papers. (The conclusion was that the cull of several MILLION animals had not been necessary). I'm thinking of writing something about it.

[LATER, SAME DATE]

I think I have solved the issue re figure 18 (it is, indeed, really badly written). I still think there is an issue, but it is one I understand (about model comparison). So you may ignore that request.

If you have any comments on 8.4.3. I'd still be very interested. All I know about collinearity is that the assumption of independence of variables is violated. I've never properly known what follows from that.

I'm not sure now the model itself is written in R. It's maybe just the file-fetching stuff. I think the model itself is written in python.

I've been Ransome writing all day and just spent ten minutes looking at the telly. Liberate Virginia it said. Sounds okay to me. Of course, you may die doing so.

To GRP 20 Apr 2020.

This week I bring you the US covid model. The update notes and faq sections are well worth reading. (Read right through, because it is in chronological order up to April 17.
https://covid19.healthdata.org/united-states-of-america.

If you click the country panel on the opening page there are data for many European countries. The model is really a curve-fitting exercise, but none the worse for that. It keeps its feet on the ground and doesn't extrapolate wildly. Overall, the data suggest that the first wave at least is over in France. Not quite in the UK (our NHS is set fair to have one of the worst outcomes in the world). In France, the President has indicated that things will change on May 11. What exactly will change, remains unclear. Schools are to re-open, but this seems unlikely given the teachers oppose the action while bars cinemas and restaurants are to stay closed (presumably for good reason). France is, therefore, in only slightly less of a mess than other places. Garden Centres and DIY stores are still closed and may not re-open in smaller towns, given they were only scraping along (there are two such where I live). Many small shops have closed and will probably not re-open, their owners now being bankrupt. Some street markets will possibly re-open, but many stall-holders live on the margin and may think it not worth the effort.

Since the definition of the end of the first wave is one case per million, I do wonder where the second wave will come from, but I suppose that only shows the depth of my ignorance. I am trying to discover.

This website (in contrast to the EUROMOMO site) indicates the actions taken in different countries (and different US States). This is very informative. For example, Sweden is far from the only European country that avoided draconian lockdowns. In fact, if you simply look (the eyeball test is quite useful in statistics) at the range of actions taken, lockdown (defined as "suppression" by the team at Imperial College) seems to bring no very obvious advantages, other than a slightly shorter epidemic. In this context, I am puzzled that so few countries have travel bans - you can't help wondering why? I wish I knew more about this stuff. I also wonder why so many of the Professors saying overreaction are retired/emeritus and all the Professors saying lock-em-up are not. Just asking.

Fortunately, it will be possible to do some statistics on these data soon and know, for example, whether Sweden's approach was right after all (the answer will be yes). Unfortunately, the results will be difficult to interpret because the actions taken by various countries overlap and are nested within each other (i.e. if you lock people up, it's pointless counting the banning of mass gatherings as an independent action). This is a non-trivial issue in modelling and is (only very briefly) addressed in Professor Ferguson's paper. So far as I can see, the paper doesn't weight actions at all, or at least doesn't say it does. I have read the paper (as an amateur modeller) several times. It is not at all difficult to follow (apart from its suspiciously blind-you-with-science appeal to some very obscure tests) and adopts a surprisingly simple-minded approach, given what flowed from it. In only one place is it hard to understand (the section dealing with model comparison): that is, to be charitable, very badly written. My concern is that the paper only looks at alternative models in a very half-hearted way (there is something towards the end, with a comparison with a simple alternative). The most obvious complicated alternative is what you might call the normal mortality model. Does this disease, with its differential attack by age, really differ from other diseases? (I guess most diseases have differential effects by age). The take-away message (and Prof Ferguson would surely agree) is it's notoriously difficult to compare models and show one is better than another. I wish they had spent more time on this issue, given the mess we're in.

On another front there are interesting hints that the lawyers are at last poking heads up. A Human Rights challenge (in the UK) to restricting exercise in children with learning difficulties. I believe the regulations were changed without court action (indeed, are the courts open?). Even I can see that the proposals being floated to keep the >70s locked up for ever might - very possibly - look like age discrimination. But then I'm not a lawyer.

To A 22 Apr 2020.

Sorry you're gloomy. Hope it wasn't me. If I try to analyse my emotions at the moment, I think it's more anger than sadness. Our rulers really are stupid, cowardly and selfish. Worse in the UK? Perhaps.

There's a piece in the *Telegraph* today by Philp Johnston. I'll paste it at the foot. Since he's saying what I have believed from the start, I cannot but agree. Johnson has fallen for what we call in this house a "pink hat." They used to be called Slone Rangers until they all discovered Science.

There's also a nice piece in *The Spectator* (I can't cut it, but you get three free reads) about Sweden. I believe the conclusion there is that sensible precautions (doesn't everybody over 70 take those?) led to an epidemic not worse than usual and could have done so elsewhere.

I have been trying hard to understand what this second wave story is about. It appears (at least in some measure) to be irrational. Project Fear if you like. There will be another wave largely because one has always occurred (the reasons are complicated). The claim now is that any such will coincide with seasonal 'flu. Part of the Public Health message is "get vaccinated". I usually don't unless the message in the relevant year is that the current vaccine is effective. I thought most people did that - maybe I'm wrong. If it is a vaccine against the wrong strain, you risk catching that year's right strain just when the vaccine has lowered your resistance! It will be a very hard decision this winter. I very much doubt whether an effective COVID-19 vaccine will be ready for the winter. They didn't manage one for SARS or Bird 'Flu.

There is a special kind of UK groupthink associated with the NHS and it appears to have become hysterical in London (where I gather 250,000 (!!) inhabitants have fled to their country pads). One can only hope the French take a more sensible line. I gather (from *Le Figaro*, so may not be reliable) that a Department-by-Department unlock might occur. That would, at least, be sane. It's hard to get the stats (there is a site but it crashes my computer) but there have been virtually no cases here (11 when I last looked) and not many more in our neighbouring Department (I imagined them as a bit of a hotspot, but the numbers there are also very low). I remember seeing [redacted] but I've forgotten - sorry - but I think it was also quite low. A very different story in the NE and Paris, of course.

I'm going to join a Zoom meeting in Scotland at the weekend. Geriatric academics one and all. I'll see what there is to report back.

Here are some extracts from the *Telegraph* piece. He's wasting his time, of course, politicians don't understand that the Precautionary Principle is only half a story.

"There is a growing clamour to blame Boris Johnson for being asleep at the wheel, but how fair is it?' Is there a more overused word in this crisis than unprecedented. No one could have expected this, we are told. How could any country prepare for such a disaster? Who knew? Well, the Government did, for one ...

How fair is this? Tony Blair, the former prime minister, sympathised with his successor, acknowledging that hindsight is easy and that what is important now is to make the right decisions to avoid an economic catastrophe that will far outweigh the impact of the pandemic. The Bank of England's analysis setting out the likely consequences of this lockdown must make any responsible government start preparations now for an early and managed relaxation ... To say this cannot be sustained much longer without wrecking our way of life is a statement of the obvious ...

The Government needs to be honest with people about the need to live with this virus and not pretend it will be beaten. There needs to be a grown-up debate about risks, which people take in every other walk of life, from travelling on a plane to driving a car. We know they are there but do our best to mitigate them ...

According to Prof Carl Heneghan, director of the Centre for Evidence-Based Medicine at Oxford University and a clinical epidemiologist, we passed the peak of deaths on April 8, which suggests that the mortality from Covid-19 is not much worse than it was 20 years ago. Indeed, with a much higher and older population, it is statistically less so. Arguably, it has been suppressed by the lockdown and yet two countries that have not had one, Sweden and South Korea, have fared better than the UK. Prof Heneghan says infection rates halved after the Government urged people to wash their hands and distance themselves from others on March 16. But ministers lost sight of the evidence and rushed into an unnecessary lockdown ...

To paraphrase the Prussian general Helmuth von Moltke, this plan did not survive its first contact with the politics. All the carefully argued medical and economic assessments were blown apart by a fear of being blamed for a death toll that was going to rise come what may ...

It is not the virus that is unprecedented. It is the response. ..."

To O 23 Apr 2020.

The official Rules are below. It seems to me that mowing the fields may be forbidden unless it is deemed to be "professionnelle". A farming business is professionnelle - that is what justifies their moving the sheep about, muck-spreading, mowing etc. My (current) advice is that we do not mow until the rules change in our favour. My morning reading does not suggest that will be at all soon.

30 octobre 2020 - Covid-19 : un 2e confinement national à compter du 29 octobre minuit.

Le 28 octobre 2020, le président de la République a annoncé un reconfinement sur l'ensemble du territoire national à compter du 29 octobre 2020 minuit. Un décret du 29 octobre en précise les modalités. Ce confinement est décidé alors que la France fait face à une nouvelle vague épidémique causée par le coronavirus.

Dans son allocution télévisée diffusée le 28 octobre 2020, Emmanuel Macron a annoncé le déclenchement d'un 2e confinement national. Ce reconfinement est décidé pour une durée d'au moins quatre semaines, avec une réévaluation tous les quinze jours. Les modalités du confinement sont précisées dans un décret du du 29 octobre 2020 publié au *Journal officiel* du 30 octobre.

Les mesures conformes à celle du 1er confinemen.

Comme lors du confinement déclenché en mars 2020, les mesures suivantes entrent en vigueur à compter du 30 octobre :

limitation des déplacements. Pour sortir de chez soi, une attestation de déplacement devient obligatoire. Les sorties sont autorisées pour aller travailler, se rendre à un rendez-vous médical, porter assistance à un proche, faire ses courses, accompagner son enfant à l'école, se promener à proximité de son domicile pendant une heure (pour promener son animal domestique par exemple), se rendre à une convocation judiciaire ou administrative, se rendre à un lieu d'examen ou de concours. Les déplacements entre régions sont interdits (à l'exception des retours des vacances d'automne) ;

fermeture des commerces non essentiels et des établissements recevant du public comme les bars et les restaurants ou les établissements sportifs ;

interdiction des réunions privées, en dehors du noyau familial, et des rassemblements publics ;

généralisation du télétravail quand il est possible ;

cours à distance pour les universités et les établissements de l'enseignement supérieur;

fermeture des frontières extérieures à l'Union européenne ;

les établissements de culte sont autorisés à rester ouverts mais les réunions ou les rassemblements y sont interdits. Les cérémonies funéraires peuvent y être organisées mais elles ne peuvent rassembler qu'un maximum de 30 personnes.

Les mesures qui changent par rapport au 1er confinemen.

Le 2e confinement va cependant être plus souple dans certains secteurs : les cimetières restent ouverts.

les crèches, les écoles, les collèges et les lycées restent ouverts avec des protocoles sanitaires renforcés (port du masque obligatoire à partir de 6 ans**, notamment) ;

dans les établissements d'enseignement supérieurs, les laboratoires de recherche restent ouverts aux doctorants, les bibliothèques universitaires, les services administratifs et de médecine préventive peuvent accueillir les étudiants sur rendez-vous, l'accueil des étudiants peut également être possible pour les travaux dirigés ;

le travail en usine, dans les exploitations agricoles, dans le bâtiment et les travaux publics peut continuer ;

les guichets des services publics restent ouverts ;

l'organisation d'épreuves de concours et d'examen reste possible ;

les visites en maisons de retraite et en EHPAD restent autorisées dans le strict respect des règles sanitaires ;

les espaces verts (parcs, jardins...), les plages, les plans d'eau et les lacs restent ouverts (les activités nautiques et de plaisance sont cependant interdites).

[** I can barely believe I typed this!]

To P 29 Apr 2020.

I bring you a quite sensible piece from *The Critic*. It deals mostly with the situation in the UK.
https://thecritic.co.uk/its-hurting-but-its-just-not-working.
The situation in France is a little clearer following the PM's speech yesterday. If all goes well (that should really be iff), lockdown ends on May 11 to be replaced with something less severe. There is to be a 100 km radius limit on travel, but I'm afraid I don't yet know how this is to be policed.
The questions that rattle in my head have not changed much.

(1) concerning the peculiarly political nature of this pandemic; (2) concerning the differences between countries.

To P 29 Apr 2020.

Thanks. I'll read it in bed. I have been writing for six hours today. Feeling reasonably happy that the book is over the hill. But I worry that I have no notes for two chapters. It's all in my head. Sometimes that works fine. Other times ...

Trouble is I know too much - there's no way I could stuff it all in and have a book anybody would want to read. I wish I'd stopped research reading months ago.

The govt is going to release we frogs on May 11, which is a Monday. They are already telling people not to "faire le pont" and start hopping about the weekend before. A faint hope I'd say - we have plans already. Our department has not been very badly affected, apart from one city. We will know whether we are "green" in a few days.

I must say the French have dealt with this (so far) better than the Brits. Quite a lot of political bickering, but at least there has been a plan at each stage. Whisper it gently - the French Health Service is better than the NHS.

I was quite frightened of this disease, but tried hard to be rational. That means keeping fit (I walk a km twice a day uphill half the way), keeping off the booze (necessary anyway because the book has to be finished) and lopping a few kilos off the stomach (only very light lunch and no snacks). We'll see.

Must go to bed.

To P 30 Apr 2020.

Thanks. I read it last night. It's a losing battle, isn't it? (I notice Scotland rushed to extend FOI response times - they regularly ignored them anyway). I am in a big row with [redacted] at the moment. He believes in supressing what he calls fake news. I keep pointing out that what he calls fake news always tends to be something he doesn't agree with ... I adopt a free market approach to "news" - let a hundred ideas fight it out. I can't help adding that Prof Ferguson hyping up his worst case to the media like a cheap card-sharper was fake news to the power of ten. How come people take him seriously? He has never been remotely right about anything - it's a bit like inviting Caroline Lucas to be UK PM for a month or two.

Since we live in a place where there is no mobile mast signal (they exist!) and only patchy internet service, we are well insulated from the current outbreak of mass madness. We don't watch much telly either (I'm grateful I live in France, but French telly is truly dreadful) which provides real social distance.

I built a little home cinema last year and that has proved a consolation. We have a lot of Opera DVDs. I was just starting to improve the sound (to Dolby 7) when this business started up, but it's already okay (I have to wear bluetooth headphones in any case because I'm deaf).

There is something seriously fishy about this covid business. Have you seen the excess death figures for England? Something like 10 SDs up. Vastly greater than anywhere else in the world.

Chapter 3

On Beautiful Untrue Things – May to July 2020

The tranche of email letters running into the summer of 2020, are inevitably coloured by the fact that I was working at that time on a startling paradox in the life of the children's author, Arthur Ransome. Long before he wrote *Swallows and Amazons*, Ransome spent several years in Petrograd (as it was then called), combining life as a correspondent for various left-wing newspapers with a little light espionage and taking advantage of his extraordinary access to Lenin and Trotsky. The paradox I was trying to resolve relates to his very close friendship with Karl Radek, the two even sharing accommodation for a time. Radek was a singularly unpleasant character who considered that everything of political significance in life was based on falsehood. This professional cynic – along with fellow Pole Felix Dzerzhinsky – became one of the authors of the Red Terror. How on earth, I asked myself, could someone who would go on to write quintessential English pastoral romances, find common cause with a man willing to see numberless thousands perish of cold and misery? Here is not the place to go into detail, but I believe a solution to this puzzle lies in the claim that present miseries can be tolerated if the lies which support them are - as Oscar Wilde put it in his 1891 essay on *The Decay of Lying* - artistically superior to reality.

The relevance of this dangerously prescient justification to our present covid predicament can be found in an internal memorandum written many years ago by Rex Leeper, the Head of the Political Warfare Executive in the opening months of the Second World War. Addressed to civil servants, the memorandum explains the working of an Emergency Powers Act.

> The people must feel that they are being told the truth. Distrust breeds fear much more than knowledge of reverses ... The people should be told that this is a civilians' war, or a People's War, and therefore they are to be taken into the government's confidence as never before. ... What is truth? We

must adopt a pragmatic definition. It is what it is believed to be the truth. The government would be wise therefore to tell the truth and, if a sufficient emergency arises, to tell one big, thumping lie that will then be believed.

To O 1 May 2020.

Ci-joint - un croquis. C'est mon invention (c).

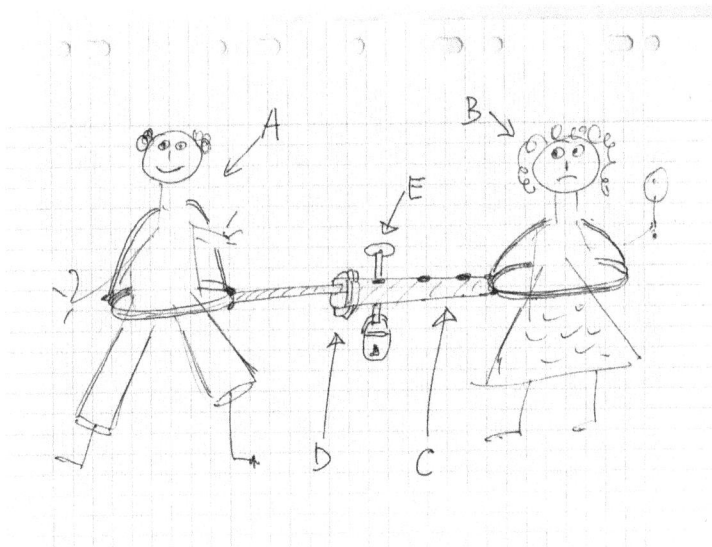

Le Séparateur

To P 4 May 2020.

I've been wondering how things wag with you. I asked and they said you were okay, so I feel able to ask without feeling I'm pestering. We are to be freed a week today. We can go out. Even mow the lawn. Speak to the neighbours. Buy some paint (at last at last) to finish decorating the palier. Take Caesar to the forest (that defines "fou de joie" ...). Poor lamb - he's been so patient confined to his ritual morning run. It's plenty of space and there's deer to chase, but he's used to change.

The Ransome book is on its penultimate chapter. There are bits missing here and there, things that I will certainly have to go back to. In places it's a mess. But for the first time this weekend I saw the end point. I'm completely off alcohol until it's finished. I dream of sitting in the garden with a plump pile of paper and glass of St Mont 2014. Come to think of it, leave the bottle.

I hope you take plenty of vitamin D. The story in France now is that the disease was here in late November. Making all those stories of the "I've had it already" kind suddenly look quite plausible. It also makes the China connection look very odd, and the unwillingness to carry out a survey of who's had it already positively sinister. You do realise it's the first seriously political disease ever. There's a paper posted in the Heterodox Academy site: three studies, all suggesting only Democrats believe in covid-19. The results are pretty amazing - have a look. We shall look back eventually and discover what this was all about.

I watch the UK telly now and then and barely recognise my former home. A kind of sloppy emotional incontinence seems to have stricken the place, last seen at the death of Princess Diana. Is it like that in Scotland? I do hope you get over it soon.

Ah well, eat your heart out .... in a few sweet short days, we shall be out and free, like that bearded guy in the *Count of Monte Christo*.

To P 9 May 2020.

Thank you for a lovely long email.

I'm trying again with attaching my invention, Le Séparateur, because I have hopes of its being adopted throughout France and the Pacific Dominions. It is now attached as a pdf. Poor French, I fear, but done in a hurry.

I'm an active member now of the Don't_Lock_Us_Up Brigade (unarmed). I may not be called to action because the Department has been declared "green." We can drive about as much as we like - even more than 100 km so long as we don't cross the County line. Which, alas, we need to do. I checked. There and back is just less than 100 km, so even a little shopping therapy in the Big Smoke will be possible. Mr Macron has locked the borders so we shall not get the usual hordes of impossibly rich Londoners driving down in their electric SUVs to patronise us in bad French at the cheese counter. You see, there is a god.

Caesar flushed out a deer this morning. I wasn't present but E said it was a sight to behold. He's now stretched out on the sofa snoring.

Hard graft on Ransome the rest of today.

To A 9 May 2020.

I should have made a distribution list - I now have more hungry mouths to feed than I expected. So, no attachments but two links instead to what look like statistically interesting analyses. Make of them what you will.

First, a new site to me. https://hectordrummond.com/2020/05/09/alistair-haimes-the-virus-that-turned-up-late/.

Second another trip to the statistician to the stars. https://wmbriggs.com/post/30606/.

There was (not much of) a pandemic. You will see Briggs also posts a link to Knut Witkowski. I read a transcript of this a while back. He is impressive.

News from France? House arrest rules are relaxed tomorrow. Somewhat.

Wash your hands if you have to.

To A 23 May 2020.

The briefest of messages just to ask is all well? I had a dream last night - I was in some kind of techno-hell waving my mobile phone about trying to hear a muffled voice that I think was yours. A day-residue dream because I had spent hours of frustration trying to get the phone to talk to the router in my office. I see people talking on mobile phones all the time. Who the hell are they talking to?

The Mayor turned up with our masks today. I asked him whether I needed to wear them about the house, trying to do irony. Oh no that's not required said he. I felt like mumbling, yet ...

Life toddles on, the grass growing at its usual rate. E asked for a DVD about Chernobyl for her birthday. Just so.

To P 25 May 2020.

1. Have you read the *Nature Human Behaviour* paper allegedly about the contribution psychology can make to the pandemic? There is virtually no psychology in it. Where it touches on areas I know about, it is wrong, which is worrying. "Psychology" seems to have disappeared to be replaced by Social Psychology opinion. Mostly very left-wing opinion at that.

2. Has he really moved? I never heard about that.

To A 25 May 2020.

Nice thought, but it's not the movie *Chernobyl* she wants, it's the film called *The Babushkas of Chernobyl* about a group of ancient women who have gone back. No bad effects (so far and it was a long time ago) and the place looks lovely, apparently. I think the film "Stalker" had the same idea. The Babushkas film isn't sold in France at all and not as a DVD in the UK (now why's that?). It is really hard to get. I found a place in Canada selling it for a small fortune.

Tell her I saw a nice "risk" statistic a few days back. A child younger than 14 has a lower risk of catching COVID than being struck by lightning. I guess for older people the odds rise to something (my favourite example) like being hit by a slate off a roof (more often than you think, but not enough to become a preoccupation). Myself, I'd dump the mask and go for the G&T. It's only Lefties and pink hats in favour of lockdown. And nice Mr Cummings, of course. But he's a fan of Bret Victor (who he, you ask? As well you might) and almost certainly Mr Trotsky.

I'm very reliably told that the race has become not to discover a vaccine (universally thought to be very hard), but to discover anything plausible at all before the virus itself goes the way of its SARS cousin and disappears. Funny old world. Good job we get another go at it.

I'll try to be better next time.

To P 2 Jun 2020.

There have been very few restrictions here for some time now (apart from bloody masks, which I refuse to wear). Restaurants and bars open today. Cinemas and theatres open next week.

Best news of all is the City dump opens tomorrow and we can start getting rid of months of accumulated garden debris.

I have been following things in the UK. The problem is the NHS. It is a state within a state. And a socialist state at that. Such things simply don't work, although humanity seems doomed to keep finding the truth of that.

I think the reaction to Covid 19 was an overreaction. As for "models" - what's the use of a model if you know nothing of the mechanism? It's become a kind of religion. I don't blame Ferguson (completely) for that - he had nothing except a model of a flu outbreak. Different disease? I assume he told somebody.

To P 13 Jun 2020.

Although it's a fairly crazy claim I have concluded that govt wants "zero carbon" above all else, and see economic planning beyond covid as an opportunity. The PM spoke in those terms (guff about a new normal and all being more generous to each other etc). Since this policy will mean breaking with the US, I don't see it getting far. Talk about green jobs and so forth is all nonsense (Spain and Germany are the compelling counter-arguments). Still, I don't live in the UK and find myself apparently drifting to the "right" in the eyes of UK friends, colleagues and relatives - whereas I see the UK as hurtling increasingly to the left. I thought the lessons had been learned. Redistribution leads to corruption and waste. Central planning leads to nepotism and stagnation. Good God, don't people remember Blair and Brown? Socialism may seem like a good idea (I was a good Methodist and I see the point), but it simply doesn't work. It's just a new set of Masters. The UK is so class-ridden it is probably beyond rescue.

Thanks for the article. As you know I was extremely sceptical about the "science" from the start. Govt surrounding itself with political enemies dressed in white coats was insane. There are umpteen reports on Imperial's previous failures. The Report on the Foot & Mouth outbreak should have meant no more funding ever. The place is a rabid nest of greenies (I think Ferguson is a Lib Dem which may be worse) with a pathological liking for the apocalypse. They think there are too many people. Little people should stop moving about and spoiling their view. Little people should "change the way they behave" (I quote). That is, preferably die or at least stop having sex. The covid hoax was obvious when the ruling class introduced exemptions for "domestic cleaners" and "nannies" in England and exceptions for "second home visits" in Scotland. And then started skipping about between their love nests. The only comfort in it all is the fact that the "vaccine" will lose the race against the disease dying out. Why the hell don't they invest their energy in understanding a mechanism for Farr's Law? If I read another paper saying "we're not curve fitting, we are modelling the known facts of epidemiology ..." I shall scream.

France (at least our bit) is back to normal. Bars and cafes are open, mask wearing is supposed to be in force but widely ignored. We were in a shop yesterday and were the only people with masks on. I had put one on (for the first time) because the sign at the entrance said you had to. The assistants thought I was being funny. I suppose I was. So I took it off.

I read Dorothy Bishop's Bartlett Lecture in the *Quarterly Journal of Experimental Psychology* today. Good topic, but lecture a flop. She doesn't know enough about the subject (the replication crisis) and kept away from Social Psychology - which was a craven decision given that's where the problem really is. So she banged on about genetics and autism - her hobby horse - but really not a stone worth lifting. I expected better given her blog.

To O 2 Jul 2020.

The postwoman (complete with bright blue mask) passed a big parcel over the gate yesterday. Well, what are we to say? What a wonderful present! Your mother has transformed a corner of our place into a very romantic place indeed. It's so extremely kind and generous. Please say a very heartfelt thank-you. It shall have pride of place.

We were thinking of you both the other evening. Dinner was BBQ salmon brochettes - it brought back memories of pre-covid days when we were allowed to go outside to play. We put paid to an excellent bottle of Medoc as the sun set.

Things are coming back to normal in France quite quickly. E wears her mask in the Post Office, but apparently very few other people do. Dogs are not required to wear masks at present. We were lucky being so isolated - although I filled in the form every morning before we took Caesar for his walk, nobody asked to see it. In fact, we saw nobody at all for three weeks! Caesar was poorly in March and E set off to buy some chicken for him. She was stopped by a Gendarme and asked where she was going, how dare she venture out blah blah. He hurried away when she said we had a vomiting invalid in the house. The uses of kennel cough.

The book about Arthur Ransome is lost somewhere in publishing hyperspace. Publishers are going bankrupt at an alarming rate. I am scared to ask. A pity, it is a nice book with some original observations. Will it ever get between hard covers? Good question.

We are avid readers of "Lockdown Sceptics" - an excellent source of sanity in a mad world. Highly recommended website even in conformist Scotia. (I suppose I shall one day be burnt at the stake, hearing cries of 'Heretic!' as my toes start to curl a little). One way or another it seems we are all doomed to be ruled by Queen Greta. At least I can say I told you so. I suppose she could always marry that Worwy fellow and start a dynasty of Wokedome.

Again very many thanks for the painting - it is really lovely. God alone knows when we shall meet again, but we shall. You know that saying, useful in hard times? 'This too shall pass...' And it will - just you see.

To P 28 Jul 2020.

I'm conscious I owe people emails - including you. I've been keeping busy, making new editions of old books and tweaking my Ransome offering.

I sort of guessed she had bitten the covid apple. It makes me angry. I sometimes feel fearful myself and that makes me yet more angry. I suppose she would consider my views unpardonable levity, but I can't help thinking the whole business a run-away laughable hoax. A bit like the CAGW hoax that was running out of steam. A transnational disease must have seemed like a manageable way of reaching "green" goals without all the slippery climate stuff. All out of control now, of course, but a hoax nonetheless. It will not end well.

As for masks - that nonsense was dreamed up by a "Young Davos Leader" named Jeremy something. Masks4all was founded a while back, complete with sinister photoshopped masked images of the people concerned. Young people in France simply won't wear them. Bravo for them. The message "you must do it to protect me" is worthy of the Nazis, as is reference to seatbelts. Believe me, I don't wear a seatbelt to protect other people! Indeed, I'm sorely tempted to start having a good stiff drink before driving off. Equally tempted to take up smoking. I fucking hate being bossed about.

I circulated my heretical thoughts on Covidery for a while, but realised I was probably causing needless offence. So I have simply continued reading round the subject, informing myself, and happily confirming I was right! Of course. Righter, in fact than I ever imagined.

A deadly disease where you need a test to discover whether you have it - that's a joke isn't it? A "completely novel" disease that turns out in fact to be an unremarkable member of a family of infectious viruses – so not "completely novel" after all. Why didn't they sack the idiot (the Chief Medical Officer, no less) who made that claim? And this bizarre concept of "asymptomatic infection?" What the hell's that? If you don't need symptoms to define a "disease", watch out, the gas ovens lie that way. And all these virgins who can infect others who may not know it either? Isn't it time that a few intelligent people ratted on this nonsense? A test of unknown specificity with a high false positive rate redefines "cases" as "infections." For god's sake - they will rise indefinitely, ticking off every one who has ever had a cold! When will people notice few get sick any more (isn't that what asymptomatic means?) and although you can never, ever, recover you no longer die. There was a nasty disease. It has run its course. Farr's Law. Even the bloody vaccine can't be tested because nobody has the disease any more that it's supposed to cure. Somebody's having a laff. I could go on.

I don't wear a mask because, although it is mandated in our bit of France, nobody here took the slightest bit of notice of the injunction (apart from supermarkets in big towns, to be avoided in any case). In any case, this is an extremely class-conscious disease and politically sensitive. We live in a very posh arty town. One believes in Covid here rather than catches it (which would be a little vulgar). Covid is a gesture of class solidarity. Mr Macron had his hopes but it was snatched from his grasp. To be clear, Covid itself only votes Lib Dem or Green or votes as a way of putting two fingers up to the bad Orange Man. Unmasked crowds are completely immune, so long as they support a woke cause, and the virus knows this. What is left of the virus only attacks right-wing thugs intending to carry out antisocial activities like singing in church and kissing people. Since notice was served on it in April, it is a virus that knows better than to attack "nannies".

Caesar send his regards. The weather here is sunny and hot and has been for several weeks. Dinner this evening will be BBQ chicken. I shall test my new three-pint starter kettle. France is nicer than Scotland.

I may be a damned heretic, but I am legally "fragile" in France and am to be protected, a bit like the tortoise or the green lizard.

To P 28 July 2020.

https://www.lefigaro.fr/sciences/covid-19-la-tres-grande-tentation-de-l-insouciance- chez-les-jeunes-20200728.
    (if you cut-and-paste, check that it doesn't introduce a space.

To P 29 July 2020.

https://wmbriggs.com/post/31960.
The mystery link was to a picture in *Le Figaro*. A crowd of people enjoying Paris café life, and going about their business. Not a mask in sight. The new PM has ordered everybody to wear masks and it is the law. But the President has already said there is no way he'll ever introduce another general "lockdown" so people are simply ignoring the mask law. It doesn't apply to the gendarmerie in any case - in itself enough for people to say forget it. It has been in the mid-thirties here every afternoon for about three weeks. Try wearing a mask in that.

The second wave is almost certainly a function of endless "testing" (the same person testing positive five times is five "cases"). Eventually, somebody will notice nobody is dying any more of a deadly disease. Briggs is right about this.

Suppression of HCQ was a big story in France. *Le Soir* covered it for days on end, although it didn't get a mention in the UK. I can't see why the people who ran the "HCQ trial" in Oxford don't end up in grave difficulty.

There are some good links on Toby Young's Lockdown Sceptics site. Perhaps you might induce her to read something there. It might reduce anxiety a little or at least raise a smile. My experience of coming out of fearfulness is that it happens quite suddenly.

Chapter 4.

La Place de l'Etoille – August to October, 2020.

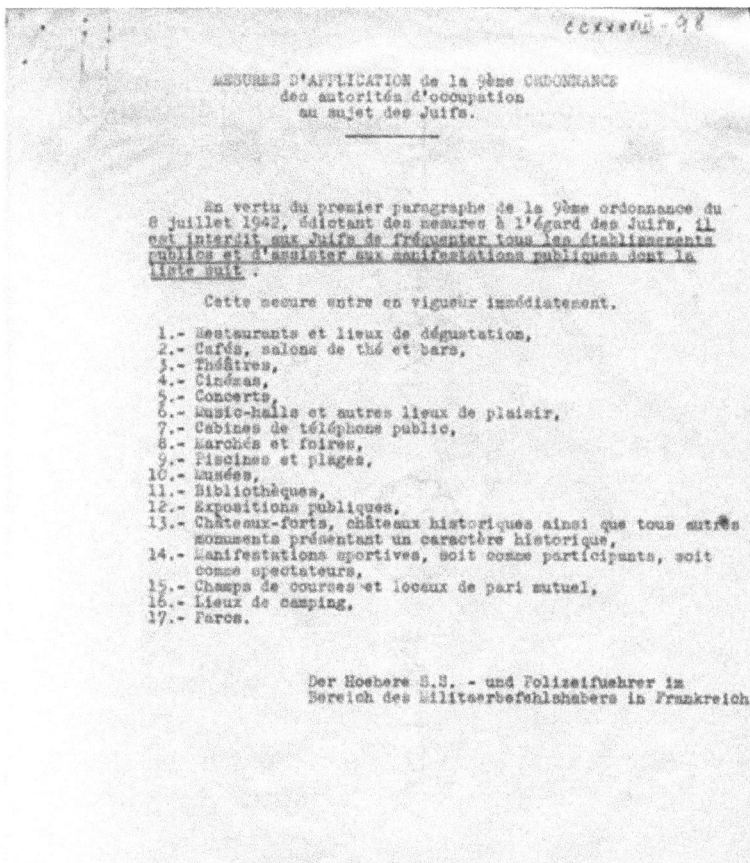

Looking through the email letters written into the autumn of 2020, I detect a sense of rising unease. The letter dated 15 August, for example, contains this comment with regard to the possibly unique position France found itself in: "Folk memory of the shameful slide into yellow star wearing is still quite potent. Some of the same arguments have been deployed regarding masks (and rapidly withdrawn).".

The reference is to the list above, issued following the application of the 9th Ordinance of the Occupying Authority "au sujet des Juifs". France, you recall, collaborated with the enemy in the application of unforgivable and arbitrary discrimination.

"How Did the World's Smartest People Fail So Miserably?" asks Toby Rogers (April 2023 -https://brownstone.org/articles/how-did-worlds-smartest-people-fail-so-miserably/). He includes Noam Chomsky on his list. Of course, you may well claim that this is very much a leading question, but Rogers is willing to provide an answer, nonetheless:

> It's over, that phase of American history, when a bunch of people baptized in the values of the 1960s, could be expected to provide the intellectual framework necessary to move society forward. There is no recovery from what they did, they collaborated with the enemy when the fate of society was on the line. To use their favourite phrase — they became "constitutive of" the predatory system they once sought to critique. Our society is so corrupt that the term "intellectual" no longer has a coherent meaning.

To P 15 august 2020.

I had intuited so. He was well when last in contact but I think my usual levity was misguided.

At least I can claim consistency. I picked this covid nonsense as the blind unfolding of "climate" (aka "end Capitalism") from the start. The same slightly barmy upper middle-class people with the same utopian objectives (no, I take back "slightly".) CAGW was always junk science. As was zero carbon (was there ever a more stupid slogan?) I see the boss of "alternative SAGE" is now talking about "zero covid". Shameless.

It's okay by me if only posh people can drive cars, fly in planes, get an education and generally live rather than exist. I'm only sorry for the really clever oiks because it will be even harder to claw a way up than it was in my day. They'll just have to work harder and keep away from universities. Meanwhile, there will be less litter on the streets.

Masks? I have only worn one once - to go into the tractor shop. The girl at the counter thought I looked funny so I took it off. I keep away from shops where you have to wear them now. Caesar doesn't wear a mask. It'll blow over in France when the kids in the night clubs get fed up. Oddly enough, although we live in deepest rural France, all the towns have huge hanger-like buildings on the outskirts, with no windows. They are packed with kids dancing and drugging every weekend. Declarations that they are closed are simply ignored. From the outside there is nothing at all to see or hear and the police know better than to try to go in.

There is a niveau of middle-class society here that doesn't exist in the UK any more. It's peopled with blokes (mostly blokes) called "intellectuals". They pop up on the telly and write opinion pieces in the newspapers. Not lightweight stuff at all, often really heavy philosophy. They are uniformly sceptical. Folk memory of the shameful slide in France into yellow star wearing is still quite potent. Some of the same arguments have been deployed regarding masks (and rapidly withdrawn).

It's hard to deny the psy-ops side of things has been pretty clever, albeit I guess it's quite easy to frighten people. But it has a fatal flaw at its heart. It's a double bind. A fearful population will lie about vaccination (which they will fear) - and I don't even count myself unusually fearful. Winding back on the fear can only be achieved by effectively admitting the thing was a kind of hoax (i.e. admitting that there was an epidemic and it ended in April). Then there is the question of culpability for the manslaughter of thousands of "care home" prisoners. When this is seen (maybe in Scotland?) as a source of party political advantage, I don't give much for BJ's chances. I assume he will flee the field wrapped in a yellow flag.

PS I gather a new Principal for a certain university is in the offing. Any news?

To P 29 August 2020.

Excess deaths? Look at Euromomo. I agree it's very odd, but both the UK and France are well below average. I don't like their switch to SDs rather than absolute numbers, but they are stuck with it. At one time I recall "England" was 20 SD above.

The data suggest the pandemic is over. Like several other 'flu outbreaks over the past twenty years. It has to be said there has not been a covid death in Scotland for a long time.

To P 30 September 2020.

I was told to arrange to hear a BBC radio production about statistics. Hurrah, at last, I thought. It was an effort, given the hour difference, but I did as I was told.

The programme was about Trump's notorious "six percent" speech (his [actually roughly correct] claim re "of" and "with"). The presenter gave the impression that he didn't like the POTUS one little bit.

They wheeled out some woman from Oxford to debunk the claim. Since "with and of" is such a subtle issue I sat back expecting something complicated and worth hearing. So much of what she actually said was so way out false, if not absurd, that I ended up continuing to think Trump had a point. He has a point.

She listed some of the "precipitating causes of death" conditions on a typical Death Certificate and called these "comorbidities". Sophistry of a high order and not funny in the context. Neither was the presenter's off-hand remark (repeated) that some people were "a little over-weight".

To be clear, in her book, the primary comorbidity in a covid death is something like respiratory failure!! It actually made me blink, but the programme sailed on merrily with no query and no comment. Where she lives apparently lots of folk are staggering about struggling to breathe. Surely every man and his dog knows the primary Covid comorbidity currently is Dementia, closely followed by BMI >35 (or some such - obesity anyway).

I can't listen to BBC radio here (it's a struggle and I shall not try again), but if that's typical, the UK is in worse trouble than I thought.

To P 12 October 2020.

The membership of the advisory group set up to (sort of) replace SAGE is still a secret. Govt say they won't tell because they are "mostly civil servants and revealing names would be inappropriate." Probably because it reports to GCHQ, which is really bizarre (or scary - take your pick) headed by the infamous Dido Harding (a one-woman disaster area). I forget its name - something with Biosecurity in the title.

I noticed that your chap was complaining he is not a member of this biosecurity setup, from which I deduce (probably incorrectly) that you-know-who isn't either. My own spies report that the govt now looks to the Alan Turing Institute for advice. That is, they are looking to AI and big data people to solve the covid problem (assuming there is one). Oddly, there don't seem to be any medics in the Institute and the psychologists are virtually all AI people (old style Sussex AI, that is). I'm sure they are all excellent chaps, but reading their bios is like taking a trip down memory lane (remember when Pat Suppes was going to solve everything with bigger computers).

Psychology as we might understand it has been completely frozen out, apart from a few token Social people, now described as behaviourists. This isn't an accident. If you go back and look at the Cummings trawl for strange people to help him, there is a list of disciplines. Neither Psychology (where exist the people who could really help) nor Linguistics (where exist all the good modelling people) are even mentioned. Our Dom seems unaware that the problems they are addressing (a) have already been addressed many times (b) are really, really, hard and (c) are unlikely to be solved any time soon. This is not going to end well.

If only there were bit more info on the workings of the Alan Turing Institute I would write something up about it.

To P 14 October 2020.

Cummings' beliefs played a part. I have read a lot of his writings. He is devoted to Bret Victor, which in my book makes him extremely naïve. But I'm not so sure about your comment on useful psychology. I don't feel rebranded - my tribe feels cast out altogether. My sort of psychologist is now seen as dangerously prone to dispute woke ideology. We're best to keep our traps shut in company. We're the awkward squad now, always saying: "actually it's more complicated than that." Need I say, nobody wants to listen? In ten years, we've become the Cassandra discipline.

No, I would not describe myself as "right wing" in the sense my father understood the term – far from it. But much of what I understand in psychology (pretty orthodox stuff) is now seen as highly politically incorrect - alt-right if you like. And that includes what I know about reading (and I know quite a lot). Okay, I'm not alone, but I do feel stranded and, in a sense, betrayed by my fellow psychologists who nipped off after the last Research Assessment to do "neuropsychology" (whatever that is) because the only price they have to pay is philosophical incoherence - and who cares about that? Frankly, I thank god I'm not still inside the university system: I would probably have been sacked by now for incorrect thinking.

Politicians make an exception for social psychologists and give them input into policy because they operate at a level divorced from psychology as we understand it (although not, of course, as they have redefined it). The "priming" scandal and multiple similar sins of commission didn't kill them off. It simply meant they now keep away from empirical research altogether. They make no statistical claims at all, indeed, they offer post-modern arguments against even trying. They don't want to know what's true, claiming only idiots look for that.

Now I accept you can't be accused of fraud if all you offer is political opinion. But the price the rest of us have paid is that hard-won, empirically based, psychological insights have been junked as well, or left to the geneticists or engineers (who are having to rediscover virtually everything we already know). It is actually painful to read computer people (the kind Hancock reveres) writing about vision as if it were something primarily concerned with data processing and optics. Or hearing social psychologists talking about reading as if it were some innate skill. Yes, I agree that eventually somebody is going to write something like *Eye and Brain* again and off we shall go again.

And, indeed, there will be soul searching - science does progress. But the truth is covid has set Psychology back decades.

To P 20 October 2020.

I think psychologists do have to bother with what (not how) they are learning. If you want to do psychology, that is. You might want to do something else, of course, like robotics. But I accept it's a deep question. E.g. I don't think language acquisition rests on the extraction of statistical regularities. It could. Indeed, it possibly should - it's just that humans apparently don't do it that way. I thought you kind of argued humans didn't do syntax ... just joking.

To P 21 October 2020.

The stats are being manipulated to produce a "second wave". I don't think the hospitals are filling up with Covid cases. Go to Euromomo. Look at Excess mortality. Add in the years they left off (left panel lets you add two years). Switch to "weekly" rather than cumulative. Feast your eyes (and notice that every "default" setting contrived to supress the truth.

There was a massive epidemic in the spring. There's nothing out of the ordinary now. Farr's Law. I guess better treatments account for something, but not much. In fact, some treatments are being denied for no good reason. Covid has simply joined the list of 'flu-like diseases that will carry all of us off eventually. The second wave is derived from current 'flu admissions and the peculiar (aka inexcusable) rules about notifiable diseases - meaning everything is classed as "covid" in hospitals.

If I'm wrong, what explains the Euromomo z scores? Below normal all-age mortality pretty well everywhere, including the UK.

To P 27 October 2020.

The problem with the death statistics is that it's been obvious for ages somebody is telling lies. On balance, I trust the sceptics - they have more to lose.

I may regret saying that when the Danish mask study is finally released. It might well say, against all the previous evidence, that there is a massive effect. My guess is it will say you can't influence a virus with a home-made cloth mask (or any other sort). Obviously, there is a risk of infection in children forced to breathe their own snot all day, but I don't know whether the study looked for adverse effects. It seems nobody will publish it until J Trump is defeated. That science should come to this.

Now ... Hoisting the flag for eye movement research (the Queen of Sciences), you don't need to correct for multiple measures if you use linear mixed effects regression. lme provides an "other things being equal" analysis free of charge, although you might have issues with the size of effects. Anyway, psychologists don't do eye movement psycholinguistics any more - it's been left to the computational linguists. I do still read in the field, but it's become very nit-picking. Now that Keith Rayner is dead there is nobody to debate with. I have declared myself the winner and not heard back yet.

Anchorite? I'd miss the booze and women. I'm thinking of becoming a social psychologist myself: you can do more or less what you like and thinking isn't necessary.

A propos, here's my first experiment. I intend to take people at random waiting outside a school and tell them their child has just been found dead. Their response to this information will define the SADE, or "Serious Adverse Data Effect" (social psychologists like acronyms). Measured SADE can then be used to quantify the nature of future "project fear" campaigns. For example, you might ensure better muzzle compliance by increasing SADE (known informally as scaring the hell out of people.) I am thinking about what the practitioners of SADE might be called - it'll come to me.

You possibly think ethical approval an issue here? Wrong - we know telling lies is perfectly justified if it gets people to "change their behaviour" [© N Ferguson]. Indeed, the more bare-faced the lie the better. This is known as "the end justifies my means" or "the precautionary principle" (take your pick). An example would be telling people they must wear a mask or they and/or their relatives literally risk dying.

A representative of the British Psychological Society sits on or near SAGE, presumably to interpret the bit in the rules about the need to avoid deception when other means are available to achieve the same ends. Whether psychologists will escape the criminal law that way could turn out to be an interesting question.

To P 28 October 2020.

It's the same here. Almost as if we are living in two worlds. Euromomo currently shows France as quite extraordinarily below the average in all-cause mortality. I think 9 SDs down, although I may be mis-remembering the exact number. It was consistent with Briggs' figure for the US. I guess prolonged national lockdowns reduce road deaths and there may be other similar artefacts in the data, but the overall picture nevertheless is of complete normality. Yet President Macron is using exactly the same phrases as Johnson (longer, flatter, more deadly etc).

The claim that covid has somehow killed 'flu is also surely false, yet 'flu deaths have disappeared. The conclusion one must draw is so obvious, it is embarrassing. How do they keep a straight face?

I am close to giving up on the issue, which consumes an hour of my day each morning. Clearly, somebody is lying. I'm also afraid the conclusion is inescapable that some people are taking advantage of a political vacuum (a reasonable definition of Johnson) to push a silly 4th industrial generation AI, Climate agenda. People who have been watching too many movies. I repeat my mantra - we'll have to get the psychology right before AI can get anywhere interesting. You saw that ad for a post doc yesterday? Somebody to work on algorithms that "learn language like humans do"? I was inclined to write and ask for a few references because I must have been asleep for a couple of decades. I seem to have missed something! Cretins.

I hope life in Scotland is tolerable. I read accounts of students being locked up and even killing themselves. It's a scandal. Why don't they just go home? I'm a very timid chap, but I'm sure I would have simply walked out by now.

To P 30 October 2020.

https://hectordrummond.com/2020/10/29/the-correlation-between-weekly-excess-deaths- and-covid-deaths-by-region-then-and-now.
(check for an additional space if you cut and paste the link).

This Hector Drummond chap seems sound. By the way, the "Prof Jones" referred to in the updates is almost certainly the Jones of Dixon and Jones (takedown of the conspiratorial ideation nonsense). He's in Physics at Oxford and seems very rational.

I broke the law this morning by taking Caesar for a walk without printing off the official form. Could have cost us 138 Euros, but we were not spotted.

Most of middle class France have shifted back to their holiday homes. Since we live in a tourist hotspot with lots of isolated houses it has vastly increased the traffic! Bloody huge electric SUVs silently sweeping past on our narrow little lanes. Still, a better class of wine in the supermarket. Stuff we rarely see because I suppose it's only worth selling in Paris/Tououse/Bordeaux. 11.5 percent old claret. And Ch Montus on the shelves. Plebs are not allowed to buy it usually. Even some Burgundy, although 50 euros seems the starting price for a decent bottle. Eat your heart out.

To O 31 October 2020.

I can only read the first screen (paywall). But it looks like it would be hard to dissent. Yeaden has a piece in the Mail. Rather poorly written but it's a surprise they published it.

Looks like another lockdown coming your way. I don't know whether to laugh or cry.

Seriously - there are very, very, clever people on SAGE (Mike Ferguson for example). Why on earth are they going along with claims that are so obviously false - "no immunity", ifr > 1 percent etc etc? I'm at a loss. And why all the stupid coloured masks?

The electoral consequences of this event (assuming elections occur) will be dire in France.

Chapter 5

The Masquerade - November 2020.

I would like to add an illustration here, but since I cannot discover who owns the rights to it, I shall have to content myself with a description. It concerns a Notice outside a Riding Club, declaring: "The Club with the Right Instincts." The attached image shows virtually everyone wearing a blue mask. This includes a little dog on a lead. Even the horse has a mask (albeit, brown rather than blue). A spare mask for visitors hangs obligingly on the fence. Emblazoned across the bottom of the notice are the stern words:

"ON FOOT A MASK IS OBLIGATORY FOR EVERYONE".

A little girl, mounted on her pony stares out of the picture, looking directly at you. It takes a few seconds to register the fact, but you eventually see she is not wearing a mask. You may well think someone is pulling your leg. Then again, read the rule: she's not on foot.

I have both a personal interest in "masks," because I am quite deaf, and a professional interest, because I have spent my academic career trying to understand the cognitive process called reading. If beginning readers cannot see the teacher's mouth, their progress will be adversely affected.

There existed in the UK a well-honed plan for what to do in a pandemic of "flu-like" disease. A question worth asking in 2020 was why wasn't it used? Did anyone ask? If so, there has not been a convincing answer yet (and no, the fact that "Covid" and 'flu" are different words won't quite do). The UK Influenza Pandemic Preparedness Strategy, in its 2011 instantiation, can easily be found on the web:

https://assets.publishing.service.gov.uk/government/uploads/system/uploads/attachment_data/file/213717/dh_131040.pdf.

It has a great deal to say about the use of facemasks and respirators, agreeing they "have a role in providing healthcare worker protection, as long as they are used correctly and in conjunction with other infection control practices." Masks (by which they don't mean bits of cloth run up on the sewing machine) provide a modest physical barrier to large droplets (sneezes and the like). You may recall the newspapers told us more than we really wanted to know about those "droplets" without really defining the term. Perhaps they thought we already knew. "Droplets" are the bits of high velocity flying snot and goo that tend to hit you in the face if somebody sneezes on you.

The Plan makes it clear that masks, of any kind, won't filter aerosols: "although there is a perception that the wearing of facemasks by the public in the community and household setting may be beneficial, there is in fact very little evidence of widespread benefit from their use in this setting." For a mask to "work" (and see above for what that means) they must be (and I quote again from The Plan) "worn correctly, changed frequently, removed properly, disposed of safely and used in combination with good respiratory, hand, and home hygiene behaviour." Now tell the truth – did you keep a mask in the car just in case? Is it possibly still there, lurking like the corpse of some expired animal festering at the back of the glove box? Did you ever keep one in your pocket? If so, why? Did you *ever* touch the thing while you were wearing it? Did you ever actually *wash* one and wear it again? (I don't believe you).

There is, in fact, a vast amount of evidence to be gained from the Cochrane Collaboration, assessing the use of face coverings during an outbreak of a 'flu-like disease." I quote: "Evidence from 14 trials on the use of masks vs. no masks was disappointing: it showed no effect in either healthcare workers or in community settings. We could also find no evidence of a difference between the N95 and other types of masks but the trials comparing the two had not been carried in aerosol-generating procedures." Importantly, it goes on to say "a mask can become dirty with excessive moisture, and contaminated with airborne pathogens. And because your voice is muffled; individuals may have to get closer to people, particularly the elderly, to hear from you.".

It was at about the time of these letters that people fell victim to a kind of mild psychological disease and starting claiming the matter was in dispute. In particular, the mask became a focus of interest for social psychologists because it was a potent symbol of compliance and a useful proxy for political allegiance. *Mask Theatre* became embedded in our social and intellectual life. It never made sense that one should wear a mask only when standing in a restaurant and only when seated in the House of Commons (and then only on one side of the room). Injunctions to wear a mask when riding a bicycle outdoors but not when riding a horse (presumably outdoors) made no sense.

To P 03 Nov 2020.

The BBC claim is false in so far as it says "UK". I go by Euromomo. All Cause Mortality was recently up a little in England (it is going down again). But it simply does not look exceptional (look for yourself). It is below the norm in Scotland, Wales and NI. My suspicion is also that any rise is outside London, where herd immunity has been reached.

To P 10 Nov 2020.

I just managed to look at Euromomo for the current available week, 44. Have a look at the SD Measure.

England +0.64 (a drop from week 43, which was 3.59. Nothing much and it's not rising, it's falling.

N.I -2.58! Wales -3.9!! Scotland +1.68 (a drop from +2.14 in week 43, both within normal range.

The context for comparison is figures like +28 for sustained periods in April, 2020.

I gather there are staff shortages in a lot of hospitals - double for the time of year. Obese staff have been encouraged to declare themselves vulnerable and shield (on full pay – what's not to like?). Also, a lot of staff have children off school after a positive pcr test (defined as an "infection").

It's not just the safety of the vaccine that bothers me. I gather the challenge to the immune system always leaves you more vulnerable to other virus infections for a time. Most vaccines don't work very well with old people anyway. We've ended up in the bizarre position of having a vaccine and nobody it would be sensible to give it to.

To A 15 Nov 2020.

Nice rant about Boris from Daniel Miller: *Who'll grab the steering wheel from out-of-control Johnson.*

> WHEN the history of the West´s collapse into collective madness and convergent opportunism is finally written, one of the most psychologically disturbing chapters will consider the career of Boris Johnson. How did the political culture of a great nation degrade to the extent that it promoted such a man to a leader? And how long will this broken personality continue to explode his psychodrama on a national scale.

> Britain´s disastrous reaction to an unremarkable disease has been shaped by the pathologies of the Prime Minister at every point. On the eve of Johnson´s election to the leadership of the most distinguished parliamentary party in the world he was living in a bedsit in south London with his pregnant mistress following the immolation of his 25-year marriage, his battered Previa GX collecting parking tickets in the street. Today, he is the animating spirit of a government combining incompetence, corruption and mendacity in equal measure.

> Panicking and hiding when the situation called for judgment and composure, Johnson´s leadership over the last 11 months echoes the general pattern of his life. The atmosphere of pseudo reality in which Britain now is frozen descends from him, as he squirms to evade accountability for the catastrophe he´s engineered, piling destruction on destruction, fiction on fiction, lie on lie.

> Why has Johnson repeatedly allowed deliberately misleading charts based on manipulated data to terrify the British public into complying with a lockdown policy that evidently could not otherwise be justified? It is impossible to think the Government does not know what we know: that the virus has an average age of mortality of 82, higher than the UK life expectancy, and an infection fatality rate of less than 0.2 per cent. Why

has the British economy been ruined and the British people terrorised for this ultimately trivial disease?

Bet he felt better after getting that off his chest.

To P 17 Nov 2020.

Just got the emails about the university rules. There is no mention of ethical approval. Do you think it was ever put to an ethics committee? I must say I find it quite sinister. Voluntary means what it says. It is incompatible with coercive efforts in the background. How can you make "going home" contingent on a test for a "disease" when you have no symptoms?

To P 19 Nov 2020.

I assume you have read the Danish Mask study. It is linked all over the place (but not the BBC), so I will not attach. Attached summary is good. I can't bother to do the calculation, but I imagine the numbers necessary to get a significant difference would be quite big. P=.3 is pretty damning. The talk is they were asked to remove reference to the downside effects (fungus infections mostly). Anyway, there isn't really an "it" to spread, is there? It's a seasonal cold. A bad one, but not worse than several in the past ten years. You can't do anything about magical thinking, can you? NB It contains the germ of an idea for a mass survey - mask wearing and political affiliation. Scope for clever examination of the shy Trump effect (hint: neck vs face).

Sorry ... dentists this morning - explains the sardonic humour. The appointment was hilarious. Massive precautions outside the surgery, spraying everything. Then the dentist did everything completely normally including using all the bloody "instruments rotatifs" - which are strictly forbidden. The ones where they keep saying tell me if it stings ... vive la France. I had a nice discussion afterwards about working with mirrors. First time I have worn a mask (I don't shop at the moment, E does it all, god bless her). You are supposed to, even in the square outside, but from casual observation I would say only about 20 percent do. Mostly old people. School kids wear them round their neck. Just the boys - girls have stopped wearing them altogether.

To P 25 Nov 2020.

I see the Spiegelhalter defence (you used it!) has perished in the face of a SAGE paper (I can send it you but it's on the govt site). In April the official line was false positive rate was unknown, but between 0.8 and 2.4. Agreement that with very low disease incidence, virtually all "cases" were false positives. Implications for any student wishing to sue for damages are interesting. Needless to remark use of the 1984 act rests on cases. This hoax could be beaten in the courts as soon as somebody can find a lawyer (they are in hiding).

To P 27 Nov 2020.

I'm not very keen on climate modelling. Too much guesswork. There's plenty of on-the-ground biology about adaptation to climate change. Nothing to suggest tipping points. I tried to make a list a couple of years ago of climate-induced extinction (I suppose that's the sort of abrupt event they mean). It's a struggle to find anything at all. The one example (I think it was a snail, but I may be mis-remembering) I wrote to John Raven about it and he thought it was indeed an example of a climate-induced extinction. Then the creature turned up alive and well in numbers. Note I am happy to agree that there are changes in the climate over time. There is no evidence these are driven by anthropogenic CO_2. None at all - which is pretty extraordinary given the investment. I follow Richard Lindzen on this specific point and he should know.

My view is that "Climate Change" is a lefty hoax, an effort to shoehorn "U-Socialism" (i.e. not Non-U) carried out by the usual elite crowd. Covid is a reflection of the same impatience that people simply refuse to vote the right way.

Re Ridley, what exactly does he get wrong? He's usually good on energy. The so-called Sceptical Science crowd (Cook et al) usually go for him, so he's normally bullet-proof.

Chapter 6

A Hot Christmas - December 2020.

These letters completed my first year as a Covid innocent. It had been a year in which, overnight, we all became amateur virologists, epidemiologists, psychologists. Informed outsiders took an interest in the defining features of a "pandemic" and those who had imagined they were going to run the show became uneasy. In blogs, websites, substacks and "alternative media" little people began asking questions they were never supposed to ask. In particular, people were noticing parallels between the response to the pandemic and the response to an alleged "climate crisis."

By Christmas 2020 more questions were being asked than our masters had ever anticipated and a lot of professional noses were put out of joint. Perhaps they should have read that folk story about letting things out of bottles.

To A 2 Dec 2020.

In bed. Lovely weather outside - pearly mist over the fields, faint yellow sun starting to break through. Hound yet to rise, no doubt (at all) curled up in front of wood burner that has kept his dreams going all night.

Thought you might like the attached. I enjoyed its energy and was particularly taken by the references to Ferguson, a man I encountered at the height of the foot and mouth pandemic (actually his boss was Anderson but scary Neil seemed to be pulling the strings). I was on RSE Council at the time. Ferguson is barmy (just my opinion). It was only a legal challenge that stopped the culling of all cattle in the UK. It turned out he knew absolutely nothing (zilch) about how the disease was transmitted. It took a group of vets to raise a court action. Everybody was unwilling to mention that NF seemed not to like meat-eaters, because it sounded so petty given the scale of what he - Nero-like - was proposing. There are quite a few other things he doesn't like, including little people and their pleasures.

I have a copy of the RSE Report if you'd like to see it. It's on the web. Pretty Mr Blair ignored it. He said he was following the science as it had been represented to him. TB was going to be EU President, you may recall. Dirty game, politics and who cares about cows?

BTW, lots of farmers killed themselves, but the NHS refused to collate the statistics - the victims being past it on admission. A job for undertakers really, they claimed. Sound familiar?

Vive la France! We are subject to regulations far, far, more severe than yours. The response has been to ignore them. I recommend it.

The annual 'flu visitation is now being reclassified as "covid". Of course, everyone predicted that would happen, but it seems so brazen one wonders whether shame can be extracted like teeth. Apart from remote rural areas (like here) the epidemic of covid is over. More than half of the deaths were old people in care homes - who simply seem to have been killed off like cattle, although not with a stun gun. (I can't really verify that bit). I don't think the bodies were burnt on the spot, but it wouldn't surprise me.

Visited a couple yesterday. Roads busy as ever. No checks of the "form". No mask-wearing zombies, indeed few muzzles left, because they get in the way of the fag. The local gendarmes (including the lovely Virginie, who surely can't be) like to hunt. They shop at the butchers and at Intermarché. Their kids go to the school. They have worked it out.

The Society of Authors has adopted guidance about "BAME" characters. Patronising tosh. Time to unsubscribe? Or fight?

To P 4 Dec 2020.

I'm in favour of vaccination, but I see the US is suggesting the Astra vaccine might not be suitable for women thinking of getting pregnant. Something to do with the spike protein target also being involved in the placenta. A bit scary that one might have that sort of concern and yet have rushed it, particularly when young women have virtually zero chance of dying from the disease. Vaccines don't work all that well on old people anyway, so I shall sit it out. If they let me. At least pro tem. Not that I'm thinking of getting pregnant. A scandal you have to do your own research simply because some doctors have turned out to be money-grubbing chancers.

To O 6 Dec 2020.

Just came into my study to see your email pop up. It is bloody cold in here. Directly above the wood burner in the room below, but that's all the heat I get. By choice I should say because I work better in the cold.

No, I don't do much cooking. We do share, but E does much more than half. I tend to do the big cook-ups of things to go into the deep freeze: sphagbol, chilli con carne, lamb curry, chicken & mushrooms etc. All good stuff, mind you. We eat well here.

Christmas has been the same with us for so long that I suppose it counts as a tradition. Presents in the morning. Caesar then gets sardines – currently one tin of Captain Cook 2000 vintage. Horribly indulgent and they are bad for him, but he adores them. We then watch his victory roll (worth watching) then put the furniture back and repair the broken bits. Afternoon - watch Traviata in the hope she recovers this time round. Weep a great deal when she doesn't. If possible now to be viewed in the "cinema" on the big screen (Ho! Ho! Ho!), but that will depend on heating the place up this year in a time of low mojo.

Dinner is always set as splendidly as we can. An excuse to polish the silver. Silly for two, but it always looks lovely and we shall look back on the photos fondly. This year: leek+pot soup (AK), Chinese prawns (EK) with E's sauce, then the best Christmas pudding we could buy (from the ex-pat section in the supermarket!) with Julia Child Creme Anglaise (AKA custard).

Later E retires to read with port while A watches *Die Hard*, also with port, like in the good old days. The first time we watched *Die Hard* it was with Alberto at Wester Alligin. Very cold winter a very long time ago. But it might even be *Red* for me this year, because I like that ginky girl's eyes. I don't like Brian Cox much so it'll be a toss up. I usually get the beast on the sofa because he prefers the wood fire.

All pretty Darby & Joan, I'm afraid.

Sorry to hear the news. I don't know about that drug. I'll look it up. Is it something widely taken, whether you have cancer or not? I must consult with my artist chum. She is very anti-drug, although her cancer battle left her a bit germ-phobic. She's made of steel, like another I know. I weaned myself off a drug once against all medical advice. On the grounds that the medical service to that point had done its best to kill me. No regrets.

To O 5 Dec 2020.

G'day, old bean. Thought I'd wish you happy birthday. Here's something not wholly uninteresting to read.
https://conservatives.global/the-church-flees-in-the-face-of-the-virus/.

Sean is a good egg.

Have you decided against a tree this year? My beloved has parked one against a wall outside, still in its green net. We'll put it up, decorate it, and no doubt burst into tears.

I saw all this misery years ago in my lonely battle against green climate zealots. It is so sad to see the inevitable conclusion already on the way. Build Back Better the idiots keep saying - they don't (yet) understand that that will include political systems. All that "voting" nonsense is bad for democracy, you see - quite apart from the litter. Just you wait. I really recommend Furedi's latest on democracy. The final chapter is a cracker. And he's right.

I'm doing a bit about the class-based attacks on Ransome at the moment. Did you notice how the crooks around Boris slipped in lockdown exemptions for "cleaners" and "nannies"? Now there are to be no travel restrictions for "high value" travellers. Why the hell do people put up with it?

Pipip and convert to ardent scepticism. If only for Christmas.

To P 6 Dec 2020.

Please circulate attached to your circle. Maybe there's a UK equivalent.

The canicule deaths, BTW were a result of the ancient (and benign) French custom in extended families of scooting off to the seaside and leaving gran to fend for herself for a couple of weeks. Alas, it was re-written by the climate crowd as the first great Global Warming Cull. I was there. Our ancient neighbour booked herself into hospital (because they have AC). It was hot (43 was the worst and that went on all night) but actually quite pleasant. I papered the hall ceiling, although E said I looked ridiculous naked.

To P 7 Dec 2020.

Open letter to sign:

https://usforthem.co.uk/open-letters/stop-masks-in-schools-now/.

To P 8 Dec 2020.

In my quest to re-establish sanity I signed up to the Great Barrington Declaration some months ago. I assume you know about it - the people concerned are all extremely respectable. I now lobby colleagues who might share some of my covid scepticism.

The authors of the GBD (plus others) have now established a website:

and have started linking to significant studies of the collateral effects of the C19 "pandemic" (the quotation marks because C19 meets only the recently-revised WHO definition of a genuine pandemic).

I will not lobby you further (honestly!) but if you share any of my concerns you might investigate this site, contact the people concerned in the UK (Gupta is in Oxford.) They might be interested in a paper from you.

I am personally interested in the (largely ignored) epidemic of suicide. I have also tried to make sense of the few studies of mask-wearing. Communicative competence (perhaps even language comprehension itself) may well demand face-to-face interaction. In which case, mask-wearing - something mandated by social psychologists largely concerned with 'control' - could turn out to have been a very bad idea.

To P 8 Dec 2020.

That's interesting, thanks. Where's the data?

I'm trying to get UK data. People seem very coy about releasing the numbers. In England (although not Scotland) the covid "switch" still applies. Suicide within 21 days of a positive test and it's a "covid death". INSANE.

Sometimes I'm tempted to add to the stats. (Only sometimes ;).

Actually, I feel quite cheery this morning, watching the vaccine story unfold. I hear the reason the priority list was changed is because resistance in the NHS itself is really huge. Probably beyond mending. The BMJ survey gives 40 percent of GPs won't take the vaccine. Since they would have been chary about declaring, it's probably much higher, they can read, after all.

It would be an interesting exercise for some psychologists (even social psychologists) to offer views on how this affair will end. I know what the agenda 21 nuts expect, but the downside at an individual level would be so adverse I think that is unlikely.

My tip is the fall of Johnson, possibly after a bad Brexit and poor election results in May. The prc scandal has to hit the newspapers eventually. Then .... anybody (literally ANYBODY) who offers an escape route will get the reins. It has to be a political solution.

Sorry to sound so banal.

To A 8 Dec 2020.

My beloved prodded me in bed yesterday with the news. I've been languishing a little just lately with a chest infection that will not quite go away.

Well, our pens have not crossed for a while. I can't remember how long ... time in this miserable state serves no purpose, one day being much like any other. I have been fighting a losing battle with depression for what seems like months. Come to think of it, it HAS been months! I shall not give in and drug myself, so I simply have to live with no alcohol, too little or too much exercise and (virtually) no sleep.

So what's new? Ransome is still on my publisher's desk. He tells me "next week" every month or so, but I have concluded they are fighting to stay alive, along with the publishing world generally. I wrote a pretty Introduction and expanded a section on P G Wodehouse, involving weeks of happy research (a very interesting man - I realised he was a teenager when Ransome's Wilde trial was on). Discovered quite a bit about Rex Leeper (more than anyone should want to know).

No, he will only communicate on his terms, and not about "it" - he thinks I am dismissive of "Our NHS" when all his chums are literally pegging out before his eyes. Tosh, of course, but his tosh. It has been very sad to see the long-planned climate plot slowly consume all in its path. I thought at one time there would be some kind of popular revolt - it seems not. I've decided to fight where I might be able to win and currently only struggle against mask mandates in schools. I think that is a battle that might be won. It's all complete nonsense, of course, but there you are.

Life toddles along. E's park of trees looked so wonderful in September, the oaks already taking shape and the grass neatly mowed in lines. It brought a tear even to my eyes. People were stopping to look. She is justly proud of it. She has transformed a whole landscape - there's not many people can say that.

I have managed to rig up a digital receiver for TNT (is it really called that?) and we can now watch French telly. The new television set is somehow so realistic, everybody looks as if they are too close to the screen. French tv seems dreadful, but there are things on Arte now and then.

I got another contract with the university in Hong Kong. Quite a lot of work. There has been a great deal of to and fro about payment. Things have changed in HK. I guess I'll get the cheque eventually. Good job we don't need the money (can you hear the hollow laugh?).

I have rejoined my French get-togethers and enjoy the fact that I can hear everybody. We are no longer Cisco Webex, which sounded very posh and was rather snappy with lots of clever ways of showing bits of film and so forth. We are now Zoom. A step down, but I guess it's cheaper. My classmates all seem depressed and weary. When asked how their week went, they all say the same - nothing to report. Life seems to have ended. I mentioned the other day I had been discussing suicide with a colleague. She retired early and was managing with a drug regime helped by a recovered interest in singing. She had joined a choir and was coming back to life. This is not a story with a good ending, so I'll stop.

E has fallen in love with a lemon tree. That's some competition.

To P 9 Dec 2020.

On T Cell immunity. There was a review in the BMJ in September on this. I recall it got some coverage in the Telegraph. September is a while ago and I have seen no clear refutation.

Here's the paper (in passing, contains good stuff against the assumptions in the Imperial models:
https://www.bmj.com/content/370/bmj.m3563.

Yet if you go to the FAQs para 3 remains absolutely unequivocal - the whole population of the world is at risk.

https://www.ecdc.europa.eu/en/covid-19/facts/questions-answers-basic-facts

You'd think they might at least temper it a bit ... or possibly get it right. It does matter.

To P 9 Dec 2020.

I used my French meet-up session to go on about this, trying to invent the term "dépossession égoiste". Give normally powerless people a little power and by God do they use it. Maybe you remember the story in *Oscar & Lucy* of the chap in the car park about to leave who sees you needing "his" place and pulls back in. We left our car in its usual place in a field a while ago and came back to find somebody had scrawled "ceci n'est pas un parking" in the grime on the back window. I assume on the assumption we had driven "too far" (it has the number plate of the next department). We went to see the chap who owns the field (he's a neighbour with a dog that seeks Caesar out now and then). He said he knew all about it - it was a new chap he'd hired. He said "god knows ... he thinks he's a policeman now...".

Thoughts on this made me consider tweaking Professor Reicher's tail and asking him if he still holds to the communitarian view of benevolent groups, all pulling together for the greater good. That's not how I see it. I wonder what he makes of protest marches, little old ladies being pinned down and "masked" forcibly. Maybe Le Bon was right, after all. He should read my Ransome book.

To A 11 Dec 2020.

Thanks for the swift reply about *A Thoroughly Mischievous Person*. I will continue to use this email address until the university sorts out a problem with its message forwarding system.

I attach an update that could perhaps usefully go into the file about this book.

To P 12 Dec 2020.

Scientists have urged people to rethink Christmas as coronavirus infection rates increase in parts of the UK, warning the country is heading towards 'disaster'. Just because people can meet up, it does not mean they should, according to Independent SAGE. The group, chaired by former Government chief scientific adviser Sir David King. He said there is a need to highlight the risks of mixing indoors." Families who decide to meet up over the five days that Covid-19 restrictions are relaxed over the festive period must be advised on how to keep their homes safe, the experts said.

Independent SAGE is also calling for a pandemic fuel allowance so people can keep their homes ventilated while at the same time turning up the heating to stay warm.

Professor Stephen Reicher, of the University of St Andrews, said: 'Right now we are heading towards disaster.'

Is the above just poorly drafted news or has SR joined David King's bunch of commies? Can't see how he can stay a SAGE advisor, can he? Just asking.

To P 12 Dec 2020.

https://extinctionrebellion.uk/2020/01/31/defence-statement-by-sir-david-king-in-support-of-five-extinction-rebellion-defendants/.

I have material on this guy going back years. He set up "alternative Sage" as soon as covid got started. Having failed to scare people over CAGW for 30 years he saw his chance, as did many others. "Covid = Climate" has been my claim from the start. The mantra Zero Covid does have a familiar ring to it ....

I know you have views about "climate change" (weird term). To be clear, I have no doubt at all about the raw data - my argument is they obviously cannot be used as evidence for CAGW. The reverse in fact - that's my point. King's defence statement is full of errors and numerous false claims, things that have been settled for years. To give just one example global warming possibly predicts fewer, not more, extreme weather events (and the IPCC evidence supports this). But fewer storms is not scary. Ditto for several other claims. Lindzen has several excellent papers on this. The CO_2 hypothesis is simply stupid. It represents an abject failure of modelling. The idea that you could have a predictive model of something as complicated as movement of a gas will eventually be seen as the scandal it was.

Incidentally that is why "climate science" is driven by social science people. Physicists find it embarrassing, as do statisticians. (I have a paper that suggests you might with a supercomputer get temporal resolution of greater than 10 MSEC for the fluid dynamics of a real gas. These climate idiots are talking about tens of years! I weep).

"Climate Change" was always going to be an effort to dismantle "capitalism" and usher in crony communism. Covid has so far proved a better way.

David King is not politically independent. Is he a tool of the CCP? It is clear that while he doesn't want all the little people to fly around the globe, he thinks the rule doesn't apply to him. Covid was the only way to kill the airline industry (the Greta Conjecture I wrote about in February) because it represented a very visible demonstration that the CO_2 hypothesis must be junk. Why otherwise exempt it in the Paris accord?

To P 12 Dec 2020.

Since you object, here's my evidence on the "Covid = Climate" claim. Some quotes from "The Green Bridge: From Tackling Coronavirus to Tackling Climate Change" by Colin Hines (Mar, 2020).

Green New Deal Group member Colin Hines makes the case for medium term efforts to tackle coronavirus to act as a 'green bridge' for tackling climate change.

One thing rarely mentioned in the wall to wall coverage of the coronavirus outbreak is how the measures taken are inadvertently helping the fight against the climate emergency. In China, the world's largest carbon emitter, there has been a 25% reduction in the carbon dioxide pumped into the air thanks to less coal burning.

Of course, the key question is how can these medium term efforts to deal with the coronavirus be built on and turned into a permanent green bridge towards tackling climate change.

We have an important window of opportunity. Major economies around the world are preparing stimulus packages. A well designed stimulus package could offer economic benefits and facilitate a turnover of energy capital which have huge benefits for the clean energy transition.

These illustrate two of the three approaches emerging for tackling the coronavirus that will also be central for tackling climate problems. Firstly, a panicked political class will continue to do what the 'experts' advise them. Secondly, the huge cost of propping up care and health systems and supporting a generalised weakening of the economy will need massive increase in government borrowing and doubtless the need to resort to coronavirus QE (quantitative easing). Thirdly, personal freedoms to do things that would otherwise make matters worse, such as travelling where we want to, bulk buying whatever we feel we need, whenever we want it, will be constrained by the State.

To P 12 Dec 2020.

And more. Quotes from Steve Reicher, psychologist, on the same theme (June 2020):
https://www.dw.com/en/coronavirus-and-climate-change-collective-action-is-the-only-way-forward/a-53595386.

If you look at the literature on what happens in emergencies, the traditional literature plays into this notion of the public as a problem — the idea that human beings are always psychologically frail and they always have difficulty in dealing with complex information. And under a crisis, they crack, they panic. You would never have a Hollywood disaster film without people running, screaming, waving their hands in the air and blocking the exits. But actually, that isn't what happens in disasters. When people come together, when they have a sense that others will support them, especially in situations of difficulty, then it makes them better able to cope and more psychologically resilient. Collectivity is the resource that allows us to cope practically, but also psychologically, to get through these times.

If you are talking about the events that are happening now due to climate change and that are killing people, it is probabilistic that climate change was critical to them. The probabilities are very, very high. But it is not immediately self-evident in the same way that it's evident that somebody is dying from coronavirus. These things become arguable. And that's where the second factor comes in, which is the political factor. In some places it has been consensual, and it has been pretty positive. And that's because politicians have not tried to argue or mobilize against compliance with medically-necessary measures. In other places, that's not true — in the United States, for instance, where Trump has been supporting those in various states who have been calling it a 'lockdown tyranny.' And in Brazil, and in India. At the moment we are acting collectively towards members of our community who are currently alive, and we can see whether they will live or die. It is much more abstract in the sense of climate change because we are acting for many of those who are not yet born — they might be our children or grandchildren.

I don't think it's entirely coincidental that some of the countries where coronavirus is raging most dangerously are those with toxic leadership, as in the United States, as in Brazil. Whereas in some of those countries which are doing well — like New Zealand — the leadership takes a very different form indeed.

The group is always going to be part of the solution. Groups can do awful things and groups can do magnificent things. The problem doesn't lie in group psychology, per se. It depends on the specific ideologies and cultures that define the groups we belong to. How inclusive or exclusive are they? What are the norms and values that define the nature of our community? Are they values of compassion or are they values of strength and domination? Not all groups are good, but that depends upon the group culture. The thing that is absolutely clear, however, is that if you get rid of groups, then you get rid of the one vehicle of change that we've actually got. If you get rid of groups, you freeze the status quo. The power of the powerless lies in their combination. I think we can wield that power for good rather than for ill.

There is a problem with the debate that's going on at the moment. Some people are telling us that coronavirus is going to change the world for the good — we're going to realize that collectivity is terrible, we're going to realize that precarity is destructive and that inequalities kill. And other people are saying, no, no, no, it's going to be completely awful — we're all going to be divided, we're going have a recession which will pit us against each other. The danger of making predictions in those forms is that it gives rise to fatalism. Either you believe it's going be awful so there is nothing you can do about it, or you believe it's going to happen anyway and therefore you don't need to do anything about it. Those were the critiques, for instance, of mechanical forms of Marxism.

I don't think there is any inevitable outcome. I'm not a prophet. If we want to move forward progressively, we've got to harness the power of the collective. We've got to understand how it's within the collective that we become agents who can actually make and change our own world.

To P 14 Dec 2020.

I do think the Christmas break will need careful news management if the UK madmen are ever going to re-impose house arrest. I guess that's why a new variant has just popped up. France doesn't get a Christmas break because nobody follows the rules anyway - the police are simply not enforcing any of the rules. Paris is heaving with riots every night, people jammed together in huge crowds. E does the shopping and I carefully fill in her form because she won't bother with it otherwise. Nobody has ever asked to see it.

To A 14 Dec 2020.

Still can't get your movie to run. Will keep trying.

Meanwhile, for all those people (I am on the list) expecting All Cause Mortality to be no higher than the last few years, the data will soon be in. Here's an interesting chart. Covid runs suicide close. And there is no increase in All Cause Mortality! End of story.

https://adapnation.io/worldwide-mortality-on-track-to-be-a-record-low/

To A 15 Dec 2020.

Here is a good (if overly optimistic) review of the Fuellmich case in Germany.
https://www.nogeoingegneria.com/news-eng/coronavirus-fraud-scandal-the-biggest-fight-has -just-begun/
There's a lot of scepticism about the chances of class action in the US raised by German lawyers. But Fuellmich seems to know what he's on about, given he was partly responsible for the VW case.

Much more interesting news re Retraction Watch (itself slowly slipping into Wokedom). The alleged "debunking" (dreadful Wikiword) of the Yeadon et al 10-point critique of Drosten et al has itself been taken apart in the Readers' Comments. Well worth reading, because the debunker replied, conceding most of the points (they boil down to primers, ct values, "dead fragments"). Rumour has it (I can't find a clear ref) that the journal, rather than retract, is seeking further peer review. Obviously opens them up to criticism of the first review (carried out in less than a day!). So I guess they will cement in the paper, but at a cost. If the cost is (e.g.) publishing ct values and WHO withdrawing stuff about primers, it will be a substantial victory.

Public Health England still will not publish ct values used in mass testing labs but is almost certainly >40. Incredible.

Link to Retraction Watch very worthwhile. Now well over 30 retracted "covid" papers.

To P 18 Dec 2020.

Replication crisis. Where is Pim Levelt? Since apparently he won't, I intend to write something about this.
https://www.researchgate.net/publication/315046675_Replicability_Crisis_in_Social_Psychology_Looking_at_the_Past_to_Find_New_Pathways_for_the_Future.

To P 20 Dec 2020.

Attached. The links won't work from within the doc, but should cut and paste. Just extracts from sensible stuff about mutation. It looks like young women (I'm not one of those) should certainly not take this vaccine. Trying to discover about enhanced effects post vaccine. Worrying. Whoever wrote this was dead right about political scientists (although I thought they didn't use that term nowadays).

If our SAGE Nudgers thought up the term "mutant strain" you must admit it was inspired.

To P 23 Dec 2020.

No restrictions here at all, APART from the fact you can't go out after 8.00. I was fourteen the last time I was told that.

I'm still not sure what we'll be eating. Guinea Fowl (is that how you spell it?) I think, but supplies have been a bit fractured just lately. I'm making the pork and veal stuffing using the never-fail Julia Child recipe. Caesar gets sardines, as ever.

Attached is your Christmas Card. It is intended to be political, but I gather I'd better say so.

To P 23 Dec 2020.

I almost signed up to the Emma Kenney et al protest, but then noticed what it said implied that unconscious social priming was kosher. My reading is that it is junk (indeed it has to be junk). Priming is on the list of nudges. I'm still waiting to see some cognitive psychologists speak up. Looks like it falls to me.

Attached file to read: The Rationalization of Unethical Research Revisionist Accounts of the Tuskegee Syphilis Study and the New Zealand "Unfortunate Experiment".htm.

To A 23 Dec 2020.

Late doing this. We got talking to a neighbour. Made a convert to the sceptic cause. Thank you very much for a long and cheery email. It's easy to become a trifle distrait just lately, is it not?

Your remark about fearing infecting others struck me hard. You should STOP that thought pronto. Just one more sick idea dreamed up by a bunch of social scientists still fighting battles about science scientists. It's my turn with the test tube, yes it is, if you don't give it me I'll scream and scream .... Please kick that idea where it belongs. People were intended to infect each other. It's a design feature, for God's sake!

Your Christmas card is attached, It is by intention a little political but not all that much. Working class sixties. That's nothing. What about working class forties? Class, Class, Class ... I sometimes think all that matters in Scotland is who you're allowed to look down on. I'm sure they have courses on it at that egalitarian paradise St Andrews? I am about to write to the pompous President of the RSE, telling her I do not share her views on the need for equality. Why is it that only the most unequal get to make that case? Don't tell me - I can guess. It's a bit like "Climate" - now chiefly the concern of paedophiles and plutocrats (what you might term intersecting sets). I want my meritocracy back. I propose telling El Presidente that I'll buy all her "let the people in" stuff, the day it applies to membership of the New Club. And will she tell me when it is going to be my turn to be President? Ouch. Must go and decorate the tree. Ho, ho, ho.

To O 24 Dec 2020.

Attached is a Christmas card. You may now print it out on suitably thick paper so that it can stand, neatly folded, next to the pipe rack over the fireplace alongside the deckle-edged invites from Bute House. I don't imagine it would then take more than a couple of hours to un-gum the printer.

Have a happy peaceful and relaxed Christmas in which nobody - not once - demands you remain safe!

To A 24 Dec 2020.

Warmest greetings from us both. We often think of you (a little enviously, given our attachment to that place). All the best for the move. Let's hope it doesn't snow! A propos, attached is your virtual Christmas Card. Nothing to scare the horses. Indeed, no animals were harmed in its making.

Many thanks for the book review. You were right about Boris Johnson. I recall asking at the time exactly why you felt so strongly he was "not right" (to quote Justine). Apart from being so obviously shallow, I didn't appreciate how serious were his flaws. That'll learn me, as my father used to say. As for the Rory Stewart piece, I ended up losing patience with it a little. It has such obviously mixed motives, trying to put himself in a good light while stabbing his old boss in the back. I wish he'd just given us the stabbing - and a lot more of it! I do recall Rory as Prisons Minister and thought him weak, given to pious phrases that came to nothing. In fact, the whole lot of them are second rate - as this covid affair has revealed.

I recall too many back-slapping events with BJ for him to be anything other than compromised. It's a bit thick complaining that Johnson wouldn't quit his FO flat - I recall him stoutly defending his boss on just that point. How they do rely on nobody remembering such things?

Thanks also for your account of these initial steps into the pit. I can't say it cheered me up - not even supper with Ms Munroe would do that – but at least it left me with a sort of melancholy afterglow.

I have an appointment to decorate the Christmas tree at 3.00, so I must go and look out some suitable music to keep us at it. Then my Christmas task: Julia Child's pork and veal stuffing. E, suitably masked, hurried off to SuperU to get the essential mixed spice. Apparently, the girl at the checkout asked what it was for!

To P 27 Dec 2020.

Re pcr test and false positives. This site provides yet more (frankly, astonishing) data on the matter.
https://probabilityandlaw.blogspot.com/2020/12/covid-19-in-uk-remarkable-divergence.html.

Good, sensible stuff. It also deals with the question of sample bias you raised a while back. I recall Spiegelhalter was called on to make the point (he's been silent since - I think he knows it doesn't hold water). Over 600 times is more than a consequence of sampling error! Remember ct>40!?

This is what these guys say about what they call causal explanations, i.e. invoking bias in sample selection: There is apparently a lot about it on twitter, but I don't do twitter.

The remarkably diverging difference between 'cases' and people reporting symptoms provides yet more evidence that the vast majority of those currently testing positive do not have the virus. There are causal explanations for part of the divergence: in March - unlike now - it was primarily patients hospitalized with the virus who were being tested, so people who may have had the virus but had minor (or no symptoms) may not have been recorded as 'cases'. It is also possible that more people with symptoms now are choosing to go straight for a test and not report their symptoms through the NHS. However, these causal explanations go nowhere near to explaining the scale of the divergence observed (625 times as many cases per people reporting symptoms since March).

To P 31 Dec 2020.

https://doi.org/10.3389/fpubh.2020.604339.
A French paper in Frontiers that reports the obvious sort of analysis that must have already been done dozens of times in the UK. I dislike the inclusion of the "climate" variable, but that is my hangup (!). In any case it seems obligatory now in French research (the more south you are the more climate money you get ... it's so bloody hot down here, you see ... ah well.).

Note the analysis can easily detect all the obvious variables – sensitivity isn't an issue. Factor 2 is the area of interest. There is no influence of lockdown policy. None at all. Whatever else masks do (and I guess it's a lot) they do not influence mortality. Ditto house arrest. Ditto stopping educating children in schools. Ditto ALL the other NPI stuff. One cannot help thinking how much all this resembles the great cattle cull of not-so-distant memory. Those days in rural Scotland you could literally smell Mr Blair's panic in the wind.

BTW, the Euromomo site now has data almost to the end of the year for France. All Cause Mortality is the only statistic not being messed about with. (Maybe I should say "yet"). Wholly consistent with the thrust of this paper. It does mean, of course, that covid will end (has ended?) with or without a vaccine. Oops.

The above paper was submitted in November, of course. I am going to ask them whether they have more recent analyses and in particular whether they have looked at LME or GLIM model construction rather than PC analysis, which will be probably be received as old fashioned.

I'm spreading this offering a little wider (using bcc, of course) in the hope you might circulate what is an important paper - obviously in its conclusions, but also in suggesting a style of analysis that gets us past the childishly naive Ferguson SIR approach, to something that is patently more complicated than that. Big data fanatics have a lot to answer for. Mind you, not half as much as Social Psychologists.

A happy New Year to all my readers. After last night I'm not sure whether it's au revoir or adieu. I'll ask Caesar.

Chapter 7

Growing scepticism - January 2021.

The new year provided (just about) enough data to assess the scale and nature of the pandemic which ran its course in 2020. Although little discussed in newspapers, the fact that the numbers of deaths from "all causes" in 2020 were not all that much higher than in 2019 was puzzling. Barely discussed at all in newspapers, informed observers were raising serious doubts about the use of the pcr test to diagnose cases of covid. Increasingly, I turned to the EUROMOMO website for its weekly updates on all-cause mortality in most EU countries (and Israel).

As a reminder, the duration of Covid-19 "confinements" in France so far had been from 17 March to 11 May, 2020 and from 30 October to 15 December, 2020. In the period covered by these letters (January 2021) the Kennedy family had just been released from house arrest, although, so far as I am aware, none of us had committed any offence. Completely innocent, we were yet to experience a third "confinement" from 3 April to 3 May, 2021.

Another flashbulb memory: on January 28, 2021, for the first time - an extremely disagreeable (set in **bold**) conclusion.

To P 2 Jan 2021.

I've come to the conclusion that all-cause mortality (in effect, the euromomo site) is all we've got by way of statistics that are at all useful. It is a complete scandal that this should be so, but there you are. The manufacture of "cases" from positive pcr tests and their morphing into "infections" is an absolute disgrace. The NHS in the UK has now saddled itself with huge numbers of nurses and doctors self-isolating, without even bothering to run confirmatory tests.

Note, that even the euromomo site is open to accusations of bias in allowing its presentation to hide the 2017 data set (the numbers are there, but you have to "add" them manually and my guess is that most casual visitors don't do this). That year (I'm tempted to say "of course") had much higher levels of all-cause mortality around the end of the year than the current data set. Context, context.

Scotland, by the way, has suddenly gone missing. Again, a little odd because Scotland had a z-score which was many SDs below average! I asked yesterday where there data are and was curtly directed to the Bulletin which just says "Christmas and New Year" (events that seem not to have affected anywhere else).

To P 4 Jan 2021.

A further critique of the Drosten paper. I have concluded there may be fraud here and that the on-going German court case will probably follow Portugal in at least making house arrest contingent on a positive pcr test illegal.
https://threadreaderapp.com/thread/1346110742924308482.htm.

This link is to Dr Simon Goddek's acerbic comments on the "two-day-review-to-acceptance" process, which he describes as a "major scientific scandal" (although he uses capital letters). Quite short and all the better for it.

In case you think me too optimistic, there has been a marked move away from mass testing in France.

To P 9 Jan 2021.

As you know, Scotland was "disappeared" from the stats in the run up to the vote on additional lockdown measures. I must have been one of many who bleated in protest.

It has now been restored. Worth looking at. It is so extremely below average that it is within an inch of being off the scale. The scale runs up to 40 sd positive and down to 20 negative (I assume because it should never go there without God's help). Scotland is currently *minus* 19.75.

Ah well.

To P 9 Jan 2021.

Can't resist quoting a comment from someone this cold and frosty morning. I wish I'd thought of it: "Coronavirus. If you go out, you can spread scepticism. People will live."

A propos, I gather the *Daily Mail* claim of "1300+ deaths in one day" (repeated in today's *Le Figaro*) is only true if you look at "when reported" rather than "when died." So, the headline is close to being a lie direct. Some of the deaths were as far back as November! Most were over Christmas and New Year. It's almost as if Public Health England were saving them up for the right psychological moment to release them ... Not possible in Scotland, of course, because there were literally no numbers. The problem in London - which is now at crisis level - is the massive number of health workers not at work, having "tested positive" and being told to self-isolate, albeit without symptoms of any disease (what we used to call healthy). As much as 20 percent in one hospital. It's a consolation to them that they are being paid, but no consolation to their patients.

To P 9 Jan 2021.

Hospitals are always full at this time of the year. The Office for National Statistics data to a few days ago suggests it was worse last year. Lots of wards are closed, beds have been removed to increase social distance.

To P 10 Jan 2021.

Provisional registered weekly death data for England/Wales for 2020 published by The Office for National Statistics. To 25th December, 62,420 deaths have respiratory illness as the underlying cause.
 By quarter :

Q1 – 22,95
Q2 – 15,83
Q3 – 11,04
Q4 – 12,57

   Now look at 2019. The Office for National Statistics data shows 71,674 deaths from respiratory diseases. 2020 is about 13% less than 2019.

Quarterly breakdown for 2019:

 Q1 – 23,33
Q2 – 16,78
Q3 – 13,46
Q4 – 18,09

   (2020 data is missing the last week.)

To P 10 Jan 2021.

LONDON **2017**:

At some emergency wards, patients wait more than 12 hours before they are tended to. Corridors are jammed with beds carrying frail and elderly patients waiting to be admitted to hospital wards. Outpatient appointments were cancelled to free up staff members, and by Wednesday morning hospitals had been ordered to postpone non-urgent surgeries until the end of the month.

Cuts to the National Health Service budget in Britain have left hospitals stretched over the winter for years, but this time a flu outbreak, colder weather and high levels of respiratory illnesses have put the NHS. under the highest strain in decades. The situation has become so dire that the head of the health service is warning that the system is overwhelmed.

Some doctors took to Twitter to vent their frustrations publicly. One complained of having to practice "battlefield medicine," while another apologized for the "3rd world conditions" caused by overcrowding.

In November, the chancellor of the Exchequer, Philip Hammond, announced an additional 350 million pounds, or $475 million, for the National Health Service in England, describing it as an "exceptional measure" to ease pressures on services during the winter. He said an additional £1.6 billion would be provided for 2018-19. But that fell far short of the £4 billion in additional funding that Simon Stevens, the chief executive of the service in England, requested last year, warning that services would come under unprecedented pressure during the winter.

"The N.H.S. waiting list will grow to five million people by 2021," Mr. Stevens said in an impassioned speech to health care leaders in November. "That is one million more people, equivalent to one in 10 of us, the highest number ever." What's more, he said, "after seven years of understandable but unprecedented constraint on the current budget, the N.H.S. can no longer do everything that is being asked of it.".

Over the past week, hospitals have increasingly declared "black alerts," an admission that they are unable to cope with demand, the health service confirmed, without releasing numbers. Most hospitals have been unable to meet emergency-ward targets of seeing patients within four hours because of a shortage of beds and staff.

Prime Minister Theresa May denied that the National Health Service was facing a crisis. "The NHS. has been better prepared for this winter than ever before, we have put extra funding in," she said on Wednesday. "There are more beds available across the system. We've reduced the number of delayed discharges of elderly people who would otherwise have been in N.H.S. beds rather than in social care," she added, referring to state-funded care. But the scenes unfolding across hospitals in Britain paint a different picture. Tuesday night, the emergency ward at Kingston Hospital in southwestern London looked more like an airport lounge than a hospital, with patients sprawled out in the waiting room.

To P 11 Jan 2021.

Attached has useful reference links on Drosten and pcr tests (my current interest in this mess). I still think legal challenges to its use in UK will occur eventually. I have no idea of the validity of the other points made, but the letter has a certain air of authority. Worth reading.

Steve Reicher and John Drury have a rather tendentious letter in the BMJ today. Since it implies "positive pcr test" = "case" = "infection" I won't give it wider circulation (the paper above is, I believe, correct in saying that transmission via people without symptoms of disease - how people used to do medicine - is probably very, very, rare - it rests on three small-sample papers from China.

To P 12 Jan 2021.

https://www.boris-johnson.com/2007/10/25/global-population-control.

It is time we had a grown-up discussion about the optimum quantity of human beings in this country and on this planet. Do we want the south-east of Britain, already the most densely populated major country in Europe, to resemble a giant suburbia? This is not, repeat not, an argument about immigration per se, since in a sense it does not matter where people come from, and with their skill and their industry, immigrants add hugely to the economy. This is a straightforward question of population, and the eventual size of the human race. All the evidence shows that we can help reduce population growth, and world poverty, by promoting literacy and female emancipation and access to birth control. Isn't it time politicians stopped being so timid, and started talking about the real number one issue?

You see the date? I wonder whether he'll come back to population control at the big Glasgow COP do? His dad was very keen on a global cull. I'm reading the new biography (kindle edition). He's a dangerous man. Carrie is the pm at the moment.

I have been really depressed today. All the misinformation and duplicity. And the sheer ignorance and stupidity gets you down. I gather between a quarter and a third of London school leavers can't read. Makes you think.

To A 13 Jan 2021.

In bed with a chest infection (again). Too little exercise is having a very bad effect on my health.

I've nearly finished reading Bower's book on Johnson. It is written in a very clumsy style, particularly the opening chapters (mostly about Johnson père) but perks up. I only have the kindle version, being unwilling to splash £20 on it. I'll wait and get a second-hand copy on abebooks - it has a good account of the rogue parliament, worth keeping for reference. It is far too kind about Oliver Letwin. To the point where one wonders ....

Bower doesn't say this, but I think it is quite clear that Johnson is and always was a "leaver" and the hostility this engendered explains virtually everything else. The wholly "remain" UK establishment still can't understand why people like me became pro-brexit. This may also explain his covid overreaction. He had been seduced into thinking that Britain can rule the Big Pharma world as the most obvious example of Global Britain. I know some of the people doing the seducing - all I can say is best of luck with that! At least I now understand why he is so keen on his big Glasgow "climate" affair. (A very odd interest for someone so consistently scathing about what he called "global warming nonsense").

Enshrining the precautionary principle into every aspect of life has always been the ambition of the climate fascists. And, a GM vaccine can be seen as instantiating the principle perfectly. That is, don't try to "treat" disease any more (old-fashioned pre-Cummings thinking), simply ensure that disease can never occur (welcome to the new technocratic paradise). Obviously barking mad, but that has never been a good argument. We can only pray the scales will fall with the departure of Carrie. If only he'd read less Horace and a little more Briggs.

In case you think I exaggerate, my university (for secret reasons) thinks you don't need biological models any more. The argument goes, what's wrong with computer models? Well, I could answer that question ...

It will sadden you to hear that I am finishing the book more pro Bojo than con. I put him on probation - he'll have to man up soon and shake off the crooks that have taken him prisoner. Lockdown as a safety-first policy is so alien to what he genuinely appears to believe, if he persists in it he will have a serious mental breakdown.

I found this below on the Lockdown Sceptics website (run by Toby Young who also runs the excellent Free Speech group - worth joining). Includes Ransome's favourite Keats quote. Anyway, it might spread a little cheer on a cold day:

I won't arise and go now, and go to Innisfree.
I'll sanitize the door-knob and make a cup of tea.
I won't go down to the sea again; I won't go out at all.
I'll wander lonely as a cloud from the kitchen to the hall.
There's a green-eyed yellow monster to the north of Kathmand.
But I shan't be seeing him just yet and nor, I think, will you.
While the dawn comes up like thunder on the road to Mandalay.
I'll make my bit of supper and eat it off a tray.
I shall not speed my bonnie boat across the seas to Sky.
Or take the rolling English road from Birmingham to Rye.
About the woodlands just right now, I am not free to g.
To see the "Keep Out" posters or the cherry hung with sno.
And no, I won't be travelling much within the realms of gol.
Or get me to Milford Haven. All that's been put on hold.
Give me your hand, I shan't request, albeit we are friend.
Nor come within a mile of you, until this trial ends.

To P 13 Jan 2021.

It's a statistical howler. Population in 2021 is about double 1940s. The all-cause mortality is up a little. About the same as 2008. The claim is discussed all over the place this morning. It is weird.

I just checked. England has recently been outside the bounds, but not by much and nothing comparable to April. It is "normal" now (week 53). Wales and NI are absolutely flat normal and have been for months. Scotland is currently *minus* 19.75 sds. Look for yourself.

The same orchestrated project fear article in the *Daily Mail* is worth reading. Whoever wrote it manages to mention the population correction well down his piece. Actually, the figures are quite interesting, albeit they do not support the case at all.

I find this perpetual diet of lies and disinformation a scandal. Where are the guardians of statistical probity? The ones who portentously took Bojo to task for his £350 million claim.

I just checked again - "Lockdown Sceptics" has a piece on the Times/BBC article. I've dumped it into a doc file. Attached.

BTW the data Euromomo present are not simply raw numbers. They are detrended to smooth out (I think this is right) consistent seasonal effects. The algorithm is published somewhere and looks sound. I recall it's a sine function. They started out simply presenting the data, then switched to this odd "sd" metric. I have a suspicion they too are doing their best to put the worst face on things. E.g. They make you fish for the 2017 data set. If you add it in it provides a sobering context for the 2021 hysteria, particularly right now (there must have been a big Dec-Jan 'flu outbreak in 2017).

Lockdowns kill. Keep the faith.

To P 18 Jan 2021.

Long-winded and rather muddled account in *The Telegraph* of points I circulated several weeks ago. Note Spiegelhalter's claims about 'flu. If "measures" have suppressed 'flu it is surely reasonable to ask why they have not suppressed covid. He also implies that non-covid deaths are running low. Doesn't say why.

To P 20 Jan 2021.

The French "SAGE" has commented recently on masks. Against the wearing of simple Cat 1 cloth masks and home-made masks and for (vastly more expensive and physically demanding) surgical masks. Very hard for this non-native speaker to intuit what they actually intend to happen. They point to the fact that certain surgical masks can be re-used, which would be an advantage (nappy washing in the 1950s style?) but add that they have to be worn correctly (e.g. not under the nose!). No change in the law yet. There would probably be a supply problem.

Site below gives a reasonable account of the situation in France at present. The interactive graphics take time to load, but are worth waiting for. Two remarks: (1) the y-axis scale varies between graphs, so check carefully. (2) there is a rare plot of stats I have never seen for UK data - numbers of "tests" (sadly type not specified) and "positivity". It seems to undermine the false positive narrative somewhat - comments appreciated.

https://www.francetvinfo.fr/sante/maladie/coronavirus/ infographies-covid-19-morts-hospitalisations-age-malades-l-evolution-de-l-epidemie-en-france-et-dans-le-monde-en-cartes-et-graphiques. htm

To P Jan 21 2021.

https://www.who.int/news/item/20-01-2021-who-information-notice-for-ivd-users-2020-0.

WHO guidance "Diagnostic testing for SARS-CoV-2" states that careful interpretation of weak positive results is needed (*1*). The cycle threshold (Ct) needed to detect virus is inversely proportional to the patient's viral load. Where test results do not correspond with the clinical presentation, a new specimen should be taken and retested using the same or different NAT technology.

WHO reminds IVD users that disease prevalence alters the predictive value of test results; as disease prevalence decreases, the risk of false positive increases (*2*). This means that the probability that a person who has a positive result (SARS-CoV-2 detected) is truly infected with SARS-CoV-2 decreases as prevalence decreases, irrespective of the claimed specificity.

Most PCR assays are indicated as an aid for diagnosis, therefore, health care providers must consider any result in combination with timing of sampling, specimen type, assay specifics, clinical observations, patient history, confirmed status of any contacts, and epidemiological information.".

This seems to be a significant change in WHO guidance. I can't see how it can possibly be implemented in the UK.

It's not being widely reported in the UK press. Maybe it's a proleptic response to the approaching German litigation over the "Drosten Test"? We'll see.

To P 26 Jan 2021.

My publisher sprang an offer on me at the weekend. If I get the Ransome MS in by 15 Feb I get into the autumn list. Lots of little things add up to my working flat out to get it done. Covid seems but a dim memory on account of my reading no news and seeing no telly. Is it still stalking the land? We received a very half-hearted invitation to consult our doctor to see whether we're suitable for the vaccine (unspecified). Like many another in France, we're waiting to see.

To P 28 Jan 2021.

It's hard to get vaccinated in our bit of France. Since I have great respect for French bioscience (far more than for UK bioscience, which seems utterly loutish and corrupt) I am tempted to a little deconstruction.

Today's *Figaro* portrays Macron with his back to the wall, pondering alternatives (continuing or extending the current curfew; regional restrictions; full lockdown). Obvious kite-flying - the situation in the Netherlands suggests there may be no strict lockdown, rather a perpetual threat of one. Meanwhile, informal action in France is undermining many of the current restrictions (many restaurants are "closed" in name only).

There is accumulating evidence that lockdowns don't work. This was, of course, the position set out in the March paper by Ferguson et al. Lockdown just spreads the disease over a longer period of time. The Imperial paper envisages on-off lockdown permanently or until there is an effective vaccine (which would be a first). Lockdowns have huge adverse consequences, which is why the WHO argued against them. I believe this message (i.e. that herd immunity is all humanity can manage) is guiding French policy behind the scenes. It is crucial to appreciate that schools are open in France and Macron doesn't want to close them, for a multitude of good reasons.

Euromomo describes the situation in France as 'mild excess' but the all-cause mortality curve is now actually below average again and has never looked alarming.

Now ... the UK. Something very, very, odd in the stats (Portugal as well??). **Horrible conclusion is these are vaccine-related deaths.** The data are being collected but will not be released until the vaccination programme is complete. Norway's published data give 0.001 adverse reaction leading to death, but this is (I think) for the whole sample, which spanned quite a wide age range. What if it is age-skewed, as seems very likely.

Three million vaccinated would be 3000 excess deaths, but are the UK excess vaccine-related deaths all in the 80+ cohort? Horrible idea, but it may account for the lukewarm French approach to mass vaccination when the disease mainly affects the very old and frail (and the vaccine simulates the disease.) You can't help noting that the second (jolt) shot has been quietly withdrawn or associated with a much longer delay. I wish to god one could simply get the numbers and see.

To P 30 Jan 2021.

Spiegelhalter has comments on deaths from covid in the UK. You need to chase his twitter feed to see them. It may appear flippant but he seems to argue that if you're going to die, you may as well do it at home and save the NHS. He has a more sophisticated way of putting it, but I am depressed nonetheless because he is a fine statistician and should know better. As predicted Macron has stalled on the question of further strict lockdown. He's up against a powerful medical establishment, but has held his ground. So far.

Chapter 8

Springtime for Social Psychology - February to April, 2021.

There was a curious omission in Dominic Cummings' famous appeal for oddball thinkers who would bring an end to Whitehall groupthink - he made no reference to psychology or to psychologists. In fact, he seemed completely unaware that the Medical Research Council had funded for many years an Applied Psychology Unit (renamed Cognition and Brain Sciences Unit in 1998) to address the very issues he was so exercised about. https://www.mrc-cbu.cam.ac.uk/history/overview/.

The kind of hard-headed empirical experimental psychology developed at the APU/CBU played a role in ensuring the safety of nuclear reactors – you might have imagined he would be interested. But no, the SAGE committee looked instead to sociologists and social psychologists.

These letters cover the spring of 2021. The dreadful days of 2020 were over and the full horror of The Global Plague had been revealed. Except that it hadn't! By then, even those most devoted to the Covid Cause must surely have realized that the response had been something of an overreaction. We had entered 2021 - the year of lockdowns, vaccines and masks – not quite understanding why any of these were necessary. We had entered a kind of cloud cuckoo land – a state described by the psychotherapist Mattias Desmet as "mass hypnosis".

Desmet is certainly correct in pointing to the abuse of the discipline of psychology in seeking to control and manipulate behaviour, but I am sceptical of his claim that loneliness and lack of social bonding led to our experiencing life as meaningless. I am suspicious of shallow appeals to the common good. What Michael Sandel, high on his Harvard perch, absurdly refers to as "the tyranny of merit" hoping to secure his argument with a snappy sound bite. You must surely be an extraordinarily sociable person to see a breakdown in social bonding as the cause of all our ills. Either that, or committed to "socialism", which is, I assume, sociability with its political hat on.

There is, in fact, something sinister in the claim made by social psychologists that human behaviour is largely "thoughtless": that our behaviour emerges from processes beyond the reach of our (rarely deployed) reflective rational mind, an idea that continues to resonate for the rather unworthy reason that it recruits its adherents into a kind of priesthood. The Sandels of this world, who claim – with a perfectly straight face – that *this applies to you, but not to me, because I, alone, see behind the curtain.* The most recent re-brand of this kind of social psychology (i.e. psychology without its mind) is the invention of *Nudge Theorists*. Persuading people to act against their own best interests by frightening them is not particularly novel, neither is it particularly ethical. Tyrants seem to know it in their cradles; prostitutes and second-hand car dealers grow into it.

Some in France (where I live) find the whole Nudge business outrageous:

Human life has been reduced to the slavery of a QR Code, with freedom granted by those of rulers in charge of vaccination. We have traded our freedoms for a spritz at the bar and divided society into the vaxxed and the unvaxxed. We have been reduced to things scanned by an app, like food at the supermarket counter. We have ended up in a state of permanent emergency, with our rulers protected from all criminal and civil liability. We are told what to do with our bodies in the name of the so-called collective. How come a vaccination is now the totem panacea for all the ills of humanity? How was this possible, in a nation such as France, in defiance of the splendour of its Constitution, its Enlightenment and the sacred principles of Liberté, Égalité and Fraternité."

[This is a very free translation from *France Soir*, because this kind of splendid stuff can only be written in French.].

In the letters dated 11 March 2021 you will find a reference to my own response:

https://www.conservativewoman.co.uk/sage-advice-built-on-fraud-and-confusion%EF%BF%BC%EF%BF%BC%EF%BF%BC/.

If you don't have time to read the lot, here's a summary of my message.

> Can we put to bed once and for all the notion that there is a mysteriously effective battery of psychological techniques, uniquely available to a new breed of psychological rulers? The grisly truth is more mundane: public compliance was secured with that same mixture of terror and virtue that Robespierre's Committee of Public Safety would have found familiar. Lies, misinformation, intimidation and (in extremis) physical violence have served tyrants for as long as history – they have no more legitimate relationship with the discipline of psychology than physical torture has with the practice of surgery. No special justification for evil acts is purchased by defining them as 'psychological' with or without an appeal to elaborate ethical codes and the alleged mysteries of unconscious persuasion.

To O 3 Feb 2021.

Yes, the document is right, as far as it goes. But the claim that "most cases of covid are asymptomatic" is very peculiar. Very peculiar indeed. Having symptoms has always been the definition of a disease. The issue of asymptomatic transmission is critical with this disease and the evidence is incredibly weak (three cases, I believe, reported in China).

I suppose you might claim that a positive pcr test with no symptoms presenting represents a "case" - but even the WHO now suggest this is wrong. The pcr test has quite a high false positive rate and the WHO guidance now suggests you need *both* a positive test *and* some kind of clinical manifestation of disease. (Otherwise I might claim I am suffering from asymptomatic ebola!) Nobody has ever proposed that "asymptomatic 'flu" even exists. It's certainly not the reason for having a 'flu jab.

Hydroxychloroquine is indeed an old drug and I don't think anybody claims it has direct antiviral action. However, used early and used together with zinc and an appropriate antibiotic it has been shown to reduce mortality. There are now trials supporting this proposition. The attempts to suppress Hydroxychloroquine medication in France and elsewhere have been quite sinister. If you are reasonably healthy and you catch the disease you may prefer to treat the symptoms (that is "be cured") or try to avoid the disease, rather than have a vaccination that does not make you immune.

Vaccination?? The trial phase of the mRNA vaccines is due to end in 2023. It's up to you whether you want to take part in the trial. It depends on how much risk you want to take. There is also a risk from the vaccine (poorly quantified at present). The data on adverse effects have not been easy to get (which itself is a bad sign).

At his age it is a tricky decision. If he's not currently meeting a lot of people and not greatly exposed to the disease, *in his shoes* I would wait for a while. Pandemics eventually die out, following Farr's Law, whatever you do. It's called herd immunity and has saved humanity for thousands of years. That is "humanity" as a whole, of course, not some particular individual.

Note - I'm not in his shoes.

To A 11 Mar 2021.

Would you be interested in using the attached article? It has not gone anywhere else for the time being.

I was unsure whether to include a biographical footnote - there is a lot of information about me on my Wikipedia page (I am the other Alan Kennedy, not the one who played for Liverpool).

With all best wishes.

To O 16 Mar 2021.

Interesting summary statistics (France) about adverse vaccine effects here.
https://www.francesoir.fr/societe-sante/deces-post-vaccination-le-droit-de-savoir.

Do they prescribe ivermectin in Poland?

To P 5 Apr 2021.

The Kennedy Family are - as you might expect - covid contrarians. Caesar as well: in fact, he knows more than most. I started a book on the great "Climate Change" scam years ago, only to discover that no one would publish it. The "Covid" scare has borrowed many of the same tactics (have you not noticed?) It is all hysterical nonsense designed to make a few people very rich and a lot more people very, very poor.

Mind you, there are some really lovely statistical howlers to delight us all (well, me anyway). One is that the risk of the "vaccine" (albeit very low) is higher than the risk of the "disease" for everybody who might be affected by the disease (otherwise known as getting old). It is a disease that you normally need a test to know whether you have it or not. Why don't people complain that sounds spooky? A disease so invisible that you are encouraged to pretend to have it. Doesn't that sound mad? A disease the cure for which makes you ill when you were not ill before and doesn't stop you catching the disease afterwards. If that were true for chickenpox, there would soon be complaints. A disease for which all the treatments except one have been ruthlessly suppressed. It all sounds like magic, doesn't it.

Above all, they claim the risk of this disease rises with age in a way no other disease has ever done - do you really believe that? Do you really believe that "covid" - whatever it is - is the only disease you can die of. Yet that is almost literally true. Are you not laughing? Next time you are shown a scary "vulnerability" graph, ask why they have not extended the age scale to cover people aged 150.

You no doubt think I exaggerate. Here's something for you to do. Look up the "adverse effects" of these vaccines. No, I'm not an antivaxxer – what I have discovered is far more interesting than that. Find the row in the tables marked "INFLUENZA". That's right - you will discover that, far from wiping out influenza, the covid "vaccine" mysteriously makes influenza reappear as an adverse effect! What's more the numbers are not small at all - last time I looked, it was a few thousands.

Following this mRNA gene therapy you can still catch the disease, still spread it, and still (if you are very old and sick) die from it. And you still (if you believe all this nonsense) have to keep away from people - who are now all defined as disease vectors. And still wear that stupid little cloth mask to declare your commitment to the cult.

By the way, since very young children can barely catch this disease at all – apparently for physiological reasons - it is probably unethical to vaccinate them, even if the risk of the vaccine is very low. I sincerely hope somebody ends up in prison for doing it.

As for the role of psychologists in this farce, you will notice they are all social psychologists. Experimental psychologists are all cowering under their blankets in case they lose their jobs. As you know, social psychologists cannot, in general, even add up without a calculator. Their understanding of probability, risk, or statistical modelling is nil - less than nil, because they hate the notion of quantitative science. .

To P 6 Apr 2021.

The UK stats have been sorted into so many different categories, it is difficult to get the overall picture. Regarding Kennedy's Influenza observation, here are a couple of clips from the current Yellow Card stats (up to mid-March, I think).

Pfizer.

Influenza 530 0.
Bronchitis 4 0.
Infectious pleural effusion 1 0.
Lower respiratory tract infection 52 7.
Pneumonia 28 6.

Astra.

Ill-defined disorder 1 0.
Illness 2732 0.
Induration 6 0.
Influenza like illness 5358 1.

The first number is cases reported, the second number is "deaths". The Yellow Card process is inherited from another age. Data are collected in rather a haphazard voluntary way, so you can assume they are under reporting, except for "deaths".

As you see, across the two vaccines there have been 14 deaths ascribed to influenza-type disease arising after vaccination. So many "serious" adverse effects would usually trigger at least a temporary stop.

As for the gross numbers, there is a category called "illness" for Astra, which seems weird to me, but since it's put in the table with "influenza" I guess you can add it.

There are over 8000 reports of "influenza" resulting from (or "following") vaccination. Even given the huge numbers of vaccinations, Influenza plainly has not disappeared.

Nobody is commenting on these figures, which I find strange and slightly alarming. Just thought you'd be interested. I do realise that vaccinations are counted in millions, but the numbers of adverse effects reported (and the way they are appearing in the press, rather than in a more formal setting) is very like what happened with the swine flu vaccines in France in 2009 - and the narcolepsy scandal that followed.

To P 28 Apr 2021.

You ask the right question - and of course I don't really have an answer. I rather despair of Psychology because it turns out to have so little to say about psychology! I wish there was a scientific hat to wear that would allow one to say something sensible about the mental world we seem condemned to experience. But as you (correctly) imply there isn't one. The fact we have to struggle on bare-headed can be quite depressing sometimes.

One amusing consequence (in UK universities, at least, I don't know about France) is that "psychology" as you and I knew it, has more or less disappeared. Apart from "social psychology" of course - whatever that is.

I have not thought about it much, but think I do believe in a concept like "archetype" and I'm encouraged in this by seeing an interview with Chomsky in which he says something like "it's worth thinking about". The more you do think about it, the less odd it seems, because we all have so much in common, including arms, legs and neural tissue. And the last has quite obvious consequences for language and thought. I suppose there is a research programme there, but it will not be pursued by psychologists in our lifetime.

As for shivering little Alan, quite so. He's still shivering. And the recurrent dream that appears in several places in my fiction is indeed one of his and not an invention. It would be boring to spell it all out, but I think I understand it. Like most people (I imagine) in the course of my life I have had one or two highly peculiar dreams, very cinematic, and very memorable. I think of them as "Jungian" dreams. I guess if I had been a patient of Jung I could have learned (expensively) to have more of them! Dreams which instantiate massive existential dilemmas.

Thanks for the encouraging words. I need them, because I have been contemplating going back to writing fiction, trying to combine my two "war" books into one. This was always my intention, but I was persuaded that the long book would simply be too long. But I'm going to have a go at it, anyway.

Chapter 9

Losing friends - May to July, 2021.

In March 2023, Scott Atlas listed what he termed a few "obvious epidemiological falsehoods" about Covid. Things that have been known to be false for years or, in some cases, for generations. Here is his list (add to it if you wish):

1. SARS-CoV-2 coronavirus has a far higher fatality rate than the flu by several orders of magnitude.
2. Everyone is at significant risk of dying from this virus.
3. No one has any immunological protection because this virus is completely new.
4. Asymptomatic people are major drivers of the spread.
5. Locking down—closing schools and businesses, confining people to their homes, stopping non-covid medical care, and eliminating travel—will stop or eliminate the virus.
6. Masks will protect everyone and stop the spread.
7. The virus is known to be naturally occurring, and claiming it originated in a lab is a conspiracy theory.
8. Teachers are at especially high risk.
9. Covid vaccines stop the spread of the infection.
10. Immune protection only comes from a vaccine.

All these assertions, albeit extremely questionable or downright false, have been recently promoted by public health officials, and academics. The letters written in the summer of 2021 touch on the psychological consequences of this promotion. They cover the dog days of 2021. Do you remember what it was like? *Something is wrong*, we felt. Badly wrong. Wrong enough for a book entitled *A State of Fear* to find readers. Wrong enough for reports of suicide. Inevitably, we deployed what psychological defences we had, hoping to ward off the worst. I write these words in 2024 and, quite evidently, we are still deploying them. The letter written on 10 July 2021 possibly suggests why.

I refer to the article by the medical journalist writing under the name "Elephant City". It is worth reading because his experiences echo my own. Asked to face reality (dare I call it *the truth of the matter*?) his parents, siblings and friends reacted with incredulity and anger, rejecting him and refusing to discuss the matter. "No argument," he says "however clever, and no data set, however telling, would change their minds." If the price of facing the truth was "the theft of their most basic freedoms", it was a price they were willing to pay. He laments the fact that only three of his closest friends resisted the Covid madness. "We've engaged in a running battle with our group about the latest abomination committed in the name of fighting Covid-19, but we've made no headway. We've finally come to the conclusion that they are beyond reach."
You can feel his anguish. How, in god's name, he seems to ask, can intelligent people believe something they know to be demonstrably false?

Well, I'm afraid the answer is *all too easily*, because minds are made that way and you are observing mechanisms of psychological defence deployed as never before. "Humankind cannot bear very much reality," said T S Eliot, who had read the works of Friedrich Nietzsche and Sigmund Freud. It's bad enough that the architects of all this insanity were so ill-educated in their own chosen domains that (by design or happenstance – it hardly matters) they got almost everything wrong - they will have to deal with that, and with their consciences, in God's good time. Much worse is the fact that this bunch of biochemical *idiots savants* may inadvertently have saddled humanity with a psychological problem of unimaginable scale and gravity. Consider this. Many of us (perhaps a majority) *know* that COVID vaccines stop the spread of the infection; we also *know* that they do no such thing. And if you think this is of no psychological consequence – just a "people thing" as one haughty medic said to me - then think again.

And believe me, it won't solve itself. To quote the same Mr Eliot: "Time is no healer: the patient is no longer here.

To O 3 May 2021.

Science has become pretty well a private affair now, conducted on blogs. I'm happy with that - it's where science used to be. If you read Montford's *The Hockey Stick Illusion* (highly recommended) you'll see how easy it was to take over the "climate" peer review process and lock sensible science out. Look at the background to Dixon and Jones for example (easily found on the web) and see how hard it was to publish a single sensible comment on a "peer reviewed" paper so bad that even I complained to the editor.

I trust Motl, Jones, Briggs etc simply because it is clear they are usually right, and when they are not, I can decide for myself. Good old-fashioned science where you show your workings. Journals are for yesterday - even in Psychology. It's obvious. Psychology is now owned by gatekeepers who get to define what you refer to as "somewhere sensible". Hmm.

I am also inclined to trust people like Hodgkinson because, although a journalist, I know he talks to people in the know. He is also retired (nobody's slave) like me! I am reliably informed that he's right about the initial response in the NHS to the reports from China. And, of course, it's well known the vaccines were patented long before then.

To P 5 May 2021.

Here is the proposed cover for my Ransome book. What do you think? I found this artist last year and suggested him to the publishers. He spent most of his life in a lunatic asylum, so I know how he feels. All the "Ransome" themes seem to be there. I see Mr Ransome as the little chap in the middle, beset by powerful women, surrounded by fairies he wished he had never invented, and facing someone with a really big axe.

I am dealing with copy editing this week. Responding to questions about my text. The woman doing this job knows more than me - which is a serious problem!

It is unusually cold here. The coldest May that I can recall, in fact.

Just in time for everybody to be wittering on about global warming. That's nearly twenty years I've been telling them it's a scam. He will soon be telling you to eat less meat - doesn't that convince you? We heat the house with wood stoves and I am still loading one every night - otherwise Caesar would object! My beloved spouse is currently planting salad vegetables outside, but would certainly send her very best wishes.

No – yet another who has ceased sensible communication. I am close to giving up on Blighty. All anyone asks is have you been jabbed? 'Jabbed' - I ask you? What kind of infantile world do these people inhabit? Bloody hell - why is NO not good enough? We frogs are not made of that kind of stuff.

To O 10 May 2021.

Sebastian Rushworth (look him up) has produced a meta-analysis of all the currently published studies on ivermectin. It is quite conclusive.

The problem (and it is a big one) is that it is virtually impossible to get hold of the drug.

My guess is that some re-packaged version will be the promised miracle drug to appear in the autumn. Once the Climate Conference decisions are made. If "vaccine hesitancy" continues, that is.

Here's an extract:

> Peer reviewed by medical experts that included three US government senior scientists and published in *The American Journal of Therapeutics*, the research is the most comprehensive review of the available data taken from clinical, in vitro, animal, and real-world studies. Led by the Front Line COVID-19 Critical Care Alliance (FLCCC), a group of medical and scientific experts reviewed published peer-reviewed studies, manuscripts, expert meta-analyses, and epidemiological analyses of regions with ivermectin distribution efforts all showing that ivermectin is an effective prophylaxis and treatment for COVID-19. "We did the work that the medical authorities failed to do, we conducted the most comprehensive review of the available data on ivermectin," said Pierre Kory, MD, President and Chief Medical Officer of the FLCCC. "We applied the gold standard to

qualify the data reviewed before concluding that ivermectin can end this pandemic." A focus of the manuscript was on the 27 controlled trials available in January 2021, 15 of which were randomised controlled trials (RCTs).

End.

To P 15 May 2021.

Are the same "stirrings" evident in the UK? There is a piece (buried) in *The Telegraph* this morning about unethical psychology. A straw in the wind?

Le Figaro today suggests tensions between a pro-lockdown "scientific establishment and a President facing tricky elections."

Although well right-of-centre, *Le Figaro* has been left of the *Daily Mail* over "Covid" for over a year.

To O 24 May 2021.

I bring you this gem from "Philosophie" Magazine, 10 December last year. Interviews with "philosophers". (Pop philosophers in France are pretty united in favour of mandatory vaccination. They are also, to a man, men – and lefty ones at that.).

I offer you a translation:

A philosopher will, without hesitation, go and get vaccinated when it's his turn. He [sic] has taken the opportunity to cast a highly critical eye on the evolution of our relationship with death, on a society weakened by exaggerated sensitivity to the natural vagaries of life, and the hypocritical cult of youth underlying it. The vaccination campaigns will restore the "delicious freedoms" of the old world: Going to the theatre and the movies, parties and restaurants, air travel across the seas, the chance to kiss and hug and shake hands. Set against these innumerable joys, a little prick, however unpleasant is nothing.

This must rank as the most stupid commentary I've seen so far.

It's about Pascal Bruckner, who I always thought was quite sensible. Mind you, I like the precious way they sneak in "air travel across the seas" (no more Orly-Nice for you, Pascal – take the bloody train).

To O 28 May 2021.

Sorry to take so long. Things Google are banned from this device. I had to copy your links and look using another computer. Nice to see you on the telly. You're a TV natural you know. Go on, I bet you rehearsed.

France is much less consumed by covid than the UK appears to be. Indeed, the only person taking it seriously appears to be our beloved President. The story here is (and always has been) the scandal of the suppression of effective treatments. I think around a half of the population has been vaccinated (mostly in the public sector). It won't rise all that much above that, whatever gets reported (the usual denominator problem). Young people simply don't watch telly any more and although there are allegedly strict regulations about going home after 9.00 (it's just gone up to 11.00) nobody bothers. And that includes me. I have only worn a mask once.

The Ransome book is currently pacing the corridors of Penguin Random House who own the rights to a lot of the stuff therein. I have someone acting for me, but his emails suggest he is even older than I am. Perhaps he's dead and I'm talking to a bot. The woman I correspond with signs herself "she/her". I'm afraid she is leading with a very glass chin with that sort of stuff, but since she has power over me, I will refrain. Pro tem.

The book *State of Fear* is available on Kindle. Excellent stuff, including some interview comments from alleged psychologists. I recommend it - only six quid. I forget the name of the woman who wrote it [Ed: Laura Dodsworth]. A journalist, but it's okay for all that.

Well, what news from here? I work at my desk four or five hours a day. I'm producing a new edition of *The Boat in the Bay*, with an autobiographical intro. And starting on the sequel to *The Things that are Lost* - nothing much, just a few cunning chapters about Laura and

Elizabeth. I know nothing at all about lesbians, so it will be a learning curve. Mind you, I've met a good few in my time.

Otherwise, I mow, spray weeds with gallons of herbicide and thrust branches into the chipper. And *water* hundreds of plants. If only global warming would really set in. I could do with a canicule.

Well, there you are old bean. De-covidise yourself. The angst has nothing to do with a virus. Honest.

To P 2 June 2021.

Hope all's well with you. Extremely (depressingly) quiet here. We go for days without seeing anybody at all apart from the postwoman (and she's started to shy away because I engage her in too much potentially infectious conversation). The fact that the Jazz Festival is virtual yet again is simply a reminder of what life used to be like here in early summer.

This tablet's battery is telling me it is only 4% alive – I know how it feels.

Ransome book in production. It looks quite pretty.

To A 26 June 2021.

Allison Pearson in *The Telegraph* about your Health Secretary. Would this even happen in France I ask myself … Here's a quote:

> Thousands of people posted reactions on social media. Some were bitterly mocking the official mantras: "Hands, Face, Back to My Place". "Saving Lives, Shagging Wives". Others were simply devastating: "I wasn't even allowed to kiss my dying father because of Hancock." The anger and disbelief were palpable. Was this really the minister who told us on the 17th May that, after fourteen months of physical and emotional self-denial, we were free to hug our loved ones, when, a fortnight earlier,

he'd been giving mouth-to-mouth to some glamorous chum he'd put on the public payroll? Knowing Hancock, he'd call it First Aide.

BTW – I remain profoundly suspicious of these periodic outbursts of conscience at *The Telegraph*, whose Covid record so far has been deplorable.

To P 4 July 2021.

A reader's comment about Psychology in one of today's UK newspapers: "SAGE may not have been all that bad, but the behavioural scientists in SPI-B were just so happy to have 66 million lab rats to control and so loathe to give up the power to deny us music, singing, companionship, closeness and even sleeping together. SPI-B need to be investigated and charged with criminal offences."

She has a point. Didn't Beckett write a play about idly waiting?

To P 10 July 2021.

I have been reading his comments about long Covid. Apparently, he's an authority on the subject (or barking mad - it's a toss-up).

So I thought you might like these thoughts from Elephant City on the loss of friends, relatives and colleagues to the Great Folly (*quant à moi, tous disparu*). Elephant City has a substack well worth chasing down. He writes very well. He's a US medical journalist.

QUOTE:

'Only three of my close friends have resisted the Covid narrative. We've engaged in a running battle with our group about the latest abomination committed in the name of fighting Covid-19, but we've made no headway. We've finally come to the conclusion that no argument, however clever, and no data set, however telling, would change their minds. They are beyond reach.'

BTW I have concluded that the ivermectin story may (eventually) make a difference. France is descending into insanity at the moment.

To O 22 July 2021.

The *Pass Sanitaire* (not simply the *Pass Vaccinale*) has been voted through. It doesn't at the moment involve supermarkets, but that may change. A lot depends on the evolving situation in England, where it is – so they say - required. Is that right?

There is - at the moment - a time limit of 30 September for the current law. It will be extended "if the situation demands."

We shall be quietly getting in such dry goods and tins as needed to see us through, although dog food is a worry. France has ways of dealing with these injunctions, but it's also a worry.

I think there is a demo against it planned for the weekend in France.

Chapter 10

On Knowing and not Knowing – August to December 2021.

On May 24 2024 Carl Heneghan and Tom Jefferson used their "Trust the Evidence" substack to summarise the contents of the package inserts that arrived along with the Comirnaty vaccine. The package inserts are demanded by the major regulatory authority in the UK, the MHRA, and represent the claims that the manufacturer is willing to make about the product they are selling.

The UK package insert claims that the vaccine is "for active immunisation to prevent COVID-19 caused by SARS-CoV-2." Active immunity means "when exposure to a disease organism triggers the immune system to produce antibodies to that disease.".

Note, "triggering the immune system" is not necessarily the same as preventing the disease. Importantly, however, there are no claims about transmission or post-exposure prophylaxis in either the package inserts or the *Summary of Product Characteristics*. The same is true of the package inserts authorised by the regulatory authorities in the USA, EU and Canada. None make any claims about impact on the onward transmission of disease.

These letters were written as the French President could be found scolding his citizens in somewhat less than presidential language, declaring that the non-vaccinated "really piss him off". Yet surely, he was well aware of the contents of those package inserts. And the Queen of England at about the same time – and no doubt equally well informed – was chastising her non-vaccinated subjects as "selfish", a word than can carry only one interpretation in this troubled context. To say nothing of a former Prime Minister characterising as "idiots" those who might have informed themselves and considered hesitation appropriate. Since he must also have been well aware that those producing the vaccines made no claims at all about vaccine influences on onward transmission, what justified the remark?

And finally, what are we to make of the literally hundreds of journal articles extolling the virtues of this or that "persuasive messaging to increase COVID-19 vaccine uptake." For example: https://pubmed.ncbi.nlm.nih.gov/34774363/.

It is surely worth asking what belief led James, Bokemper, Gerber, Omer and Huber to declare that "widespread vaccination remains the best option for controlling the spread of COVID-19 and ending the pandemic."

There is plainly a problem here.

To P 14 Aug 2021.

You're probably right. But since I'd dug it out for you, I'll send it. Professor Richard Lindzen on CO2, Venus, Mars and lots else. I have very many of his papers and even (a few years back) exchanged emails with him. He is incredibly well-informed and generally opposed by people who know a lot less about the subject than he does (including some who know nothing at all).

"Lockdown" protests in Paris quite spectacular today. I fear things will get worse: the French tend to settle matters on the street.

We are being lied to.

To O 15 Sept 2021.

Interesting piece in "The Conservative Woman", allegedly "exposing the many scandals of the Innova lateral flow test (LFT) and the US and Chinese companies behind it: Innova Medical Group, a wholly-owned subsidiary of Pasaca Capital and its Chinese manufacturing partner, Xiamen Biotime Biotechnology.

I'll quote a bit:

> The golden child of lateral tests was fast-tracked through Public Health England's Porton Down laboratory and Oxford University's joint 'rigorous'

evaluation programme. It was given special VIP status, by being placed "on top of the pile". In December 2020, it was granted Exceptional Use Authorisation (EUA) by the Medicines and Healthcare Products Regulatory Agency (MHRA), without the standard regulatory approval process because of "exceptional circumstances". Professor Tim Peto, chief investigator of the joint evaluation programme (overseen by Sir John Bell, Regius Professor at Oxford University and Susan Hopkins from PHE) admitted in relation to the Innova LFTs, "we had to buy before we knew they worked."

Hmmm. Read it for yourself.
https://www.conservativewoman.co.uk/
innovas-iniquity-part-1-how-test-firm-flashed-the-uk-taxpayers-cash/.

To P 16 Sept 2021.

I have been reading an article about, and partly by, an old Dundee colleague, Sir James Black (2011) and his crusade against computer-driven drug discovery. It possibly explains why the "Life Science" people in Dundee were so wary of him (something I've never understood, given he won the Nobel Prize!) It's a powerful argument against reductionism in Biology and the quest for single-target drugs. Can't decide whether his views were prophetic (re mRNA "vaccines") or just reflect an old chap bemoaning the inevitable march of progress.

He claims that discoveries of effective new drugs were in very steep decline in 2011 notwithstanding modelling, visualisation techniques, data mining and AI etc. Says it's because the *people* involved (i.e. not the computers) simply don't realise that Biology is extremely complicated. Sounds plausible (says another old chap).

To P 25 Sept 2021.

I've seen lots of analyses of hurricanes. In all sorts of directions. Time series analysis is black magic, and I speak as someone who did a lot of it. Since you don't know how far back the data points correlate across time (and why) you are working largely in the dark without a good model of what's going on. For example, there are n-2 effects easily shown in the eye movement record, making interpretation very hard. I can only guess how far back through a weather series you would have to go. Even the moon must be involved! The statistical model is not going to tell you what's going on - that's putting things exactly the wrong way round.

The problem is the model - showing how global warming fits in to the hurricane question - without hand waving and vague statements about more "energy in the system". Actually, I thought there was one in this case. The last I read of it, the big picture was that the differential equator-pole temperature would reduce with global warming, predicting a decrease in hurricanes intensity and a probable decrease in their frequency. It was in the last but one IPCC Report. So far as I know that's still the IPCC line, although I didn't read their last report because I think it is now all political/economic and has lost all contact with physical science. Have they modified the theory to fit the data? In which case, well done - they usually do the reverse! But it was a big thing to change, just like that.

What the global warming people need to demonstrate is that there has been a significant increase in "global" temperature (land, sea or both) over some period of time that we can all agree on (rather than continually redefining it, and pitching it forward as a modelled projection). Picking the one that Hansen used, there simply hasn't been an increase. He was wrong. There's no shame in that - no need to call him names. That's standard science - a clear hypothesis, reasonable test, hypothesis falsified. End of story. No part of New York is under water.

For myself, I'd suggest the Central England Temperature Record as a more reasonable period of time. It's not much warmer now than a hundred years ago in central England. I've not seen a convincing statistical analysis that says otherwise. Talking about average increases of fractions of a degree in a measure of "anomaly" is either barmy or deliberately deceptive. It was 11 degrees here yesterday morning when we took Caesar for a walk at about 8 am. It was 27 degrees at 1600. And that's not unusual. Even if the average here drifts higher over time by 1/10 degree or even half a degree (and there is no evidence it has in SW

France), so what? And given there is agreement that the A component of CAGW is, in any case, much less than 10 percent of the total GW, so what again? That's no reason for denying a large part of the world access to cheap energy and allowing a bunch of crooks to rip us all off selling green fantasies like "renewable energy" and "sustainable living". As with vaccines, *cui bono?* is the apposite question.

We read now (yet again) about acidification, sea rise, bush fires, wind speed, coral bleaching, species extinction and so on and so on – because that's all that's left to people - straws to clutch at. These are all putative (speculative) second-order effects in a chaotic system where the alleged first-order effect - even after 50 years of effort - is still not evident. If my side of the case was obviously wrong (in the physics sense of "wrong") you know perfectly well I'd be the first to sign up. I believe in rationality.

As for finding bias everywhere, what do you expect after reading the literature? What do you expect after the deluge of lies around Covid? What else, faced with a branch of science prosecuted by people very largely with no qualifications at all in the relevant subjects? Why is climate science a branch of Geography and not of physics? Why is Covid policy suddenly the province of social psychology? Why do real statisticians run a mile if you ask them to look at the data? Is it not more than strange that there were very few virologists on SAGE? And that the only University Department devoted to evidence-based medicine (Oxford, no less) was shut out and its boss told to shut up. In that context you'd be mad not to be sceptical.

There are books written on the qualifications of IPCC committee members. I can lend you one. Greenpeace and WWF dominate. Even the head honcho of the UN Climate Committee himself has explained that climate change has nothing to do with environmental science and everything to do with redistributing the world's wealth. I can find you the quote if you like - Stalin had the same idea. It won't end well.

I recommend the section in J Williams' book about academic freedom in the Academy (about page 65) where she discusses the use of the term "denier" in the context of the death of science as a search for truth. Climate truth is not something it is wrong to deny, it is not some kind of religious orthodoxy, whatever the BBC think. It has to live in a sceptical world or die. In my opinion, the latter, but in any case, why not allow contrary hypotheses to be assessed and weighed in the balance? I understand that classical high energy physics experiments usually look for 8 - 10 SD effects before getting at all excited. Put that alongside Phil Jones' remark at the time of Climategate that the temperature was "almost significantly higher" (that is, he'd done a t-test and got p = 0.06!). That was the chap in charge, Michael Mann's chum - the one who said he wasn't sure how you plotted a graph in excel.

To P 29 Sept 2021.

Actually, I had bookmarked a paper in Judith Curry's blog that did some analysis on CET. Google "Judith Curry the rise and fall of central England temperature" I believe there was an update, confirming the on-going local "decline" but I can't find it. Anyway, alarm bells start ringing when people begin correcting the record. That way lies the skullduggery you got in Australia and New Zealand over the same issue.

I have a great regard for Curry and corresponded once with her about climate sensitivity (when interest seemed justified and before the deluge of politically motivated "climate crap"). I was using one of the linear regression models in R to model saccade control and wondered whether one could demonstrate - statistically - the variance explained by variation in CO_2 in a given model using the same procedure ("climate sensitivity" is a crude way of coming at this). Apparently Nic Lewis had done something along those lines. Published, if that matters. What's interesting about Linear Mixed Effects models is that hugely "significant" beta weights (e.g. the linear effect of launch site (n-1) on fixation duration (n) - "all other things being equal") in the real world actually reflect tiny, tiny (indeed *minute*) differences (estimates of much less than a msec.) Frankly, they are so small, the sensible thing to do is to find something else to do with your time! Now, pick one of dozens of input variables for your standard climate (Covid?) model. Recognise the style of argument? And I'm restricting discussion to linear effects, when we all know there are massive non-linear relationships all over the place and for which there are no convincing statistical models at all. For example, in my field, the idea that fixation duration on word n is controlled by launch site from word n-1 can be sustained indefinitely. Indeed, it is "true" and you can become a launch position (or mask) zealot. I can find you papers where the "effect" is HUGE (indeed, I published some of them). But in the real world, it is nonsense. The length of the word you are looking at determines (massively) the time you spend looking at it. In the same way that the sun determines how much "energy" there is in the system. I suspect the sun zealots are going to win the day in the climate debate.

As you know, Curry eventually gave up because of all the aggro, resigning her post. I think she maintains her blog, but it's been ages since I looked at it.

We live in troubled times.

To P 03 Oct 2021.

Apparently "serious" journals (including *Nature* for God's sake!) are now publishing an alarming number (hundreds) of randomly generated fake papers. There's a nice piece in *The Daily Sceptic* about it. How, I wonder, could a paper with the title "Characteristics of heavy metal pollutants in groundwater based on fuzzy decision making and the effect of aerobic exercise on teenagers." actually get published? Don't journals even employ copy editors any more?

To P 04 October 2021.

Yes, I've published in *Nature*. It's a few years back, but I recall responding to referees (sometimes several times) as a big part of the game and very challenging. I wonder whether it has changed, though? Things got much worse when social psychologists (in particular) started pal reviewing for academic jobs - why would they not do the same for papers? Didn't Springer take over *Nature*? It isn't what it was - I prefer Climate Audit (that's a joke, perhaps).

I got my *Quarterly Journal of Experimental Psychology* this morning. Yet again, most of the papers were from non-UK universities. Some interesting things, but you don't get a sense of a clash of big ideas, big theories. Makes you wonder where Psychology is being done. Perhaps we've been killed off and I hadn't realised we're dead. Like Sociology in the 70s. A pity.

To A 23 Oct 2021.

I'm sorry to be so brief last time thanking you for reading and commenting on Ransome. I have been a bit set about with fever trees since I (foolishly) started to clear out nearly ten years of bookish debris from my study. A horrible task, made the worse by unearthing this and that, and ending up reading things. Two letters from people now dead and a letter from someone who may as well be. I do begin to feel like that growing boy with shades on all sides.

Two really huge boxes were duly poured into the yellow bins at the Corporation Dump only an hour ago. The sun was shining and the woman in charge friendly, speaking extreeeemly slow French as to a mentally deficient child: you ... walk ... to .... the ... big ... bin ... and ... you ... lift ... the ... lid. Lovely.

Any news here? Well, a bit. An interview for the Dundee Psychology Society, to be organised by a formidable young woman. I suspect she's American - they usually end up running things. The National (UK) Psych Soc apparently also want to interview me, but that's only on her say-so so we'll see. I've ordered some swanky bluetooth earbuds just in case.

I hesitate to write about Covid and have avoided writing about "vaccination" since it first hove onto the scene because of the published results of the trials. In my age bracket a one percent advantage was obviously "not nothing" (my French chum's favourite phrase). But frankly it might as well have been, given the absolutely obvious risks. Anyway, it got up my nose that this one percent was reported as "95 percent" - and continues to be. It's the old relative vs absolute find-the-lady trick. You have two samples of people, n=100 in each: only 1 catches the bug in the "vaccine group", whereas 2 catch it in the "placebo" group (usually "vaccinated" with something else). Okay, a *huge* relative difference - "shall we say 100 percent, ladies and gents? And for my next trick" .... but ... but ... but, there's still only 1 percent in it. And it's never without risk - that's the point. And don't think I am making this up - look at the figures for the clinical trials (incidentally with no pregnant "people" at all in samples of tens of thousands and only a very few oldies). The vaccinated / unvaccinated difference was less than 2 percent (no difference that I can recall in the "Old" groups). Bear in mind when the dust settles, they are going to say, with perfect justification, "we published the results ... not our fault if you misinterpreted them ... now would like a nice game of poker ... honest 'gov ... fresh pack of cards ... what've you got to worry about?".

I gather there's shortly to be a less toxic vaccine available in France. Attenuated virus technology, a bit like the classic 'flu vaccine (of course the 'flu vaccine is usually of no use at all when you're over 70!). People must either go with that or go without a social life. It appears they resent that (very much), but in fact social intercourse had stopped anyway following the UK-induced "lockdown" and the Climate-induced assault on flying. Depressed? Moi? Since when has rationality been so out of favour?

I still don't buy the Climate Crisis, by the way. Although hearing that Prince Charles had converted one of his Aston Martin's to eco fuel almost swayed me. If only he'd done both of them ... who knows? I wonder whether thoughts of his mum's demise makes his neck itch?

I have been reading the transcript of a Peter McCullough speech. He's certainly no fool, has extraordinary connections, and knows what he's talking about. The similarities with the narcolepsy scandal are all too painfully obvious. When the parents of dead children go to law in France, I can imagine him as a powerful witness. Troubled times.

Must get back to "filing". I need a folder called "dead mouse".

To P 25 Nov 2021.

Do you know how much India emits in total, and the average per person? The (alleged) relationship (as I am sure you well know) is logarithmic. Doubling and all that. Any (alleged) first-order effect is pretty well over and done with unless you are thinking in a geological time-frame. In that case 2050 looks silly - 5050 would be more like it. I assume that is why Prince Charles converted only one of his Aston Martins.

Seriously, the risk rational people run by sticking with this obvious scam is they become the tools of increasingly deranged eco freaks and advocates of socialist repression. Here's a test.

I wonder what our resident Covid expert's views are on "climate change"? And whether he could even define a basic term like one of the two definitions of "climate sensitivity". Or whether he's mugged up on black body radiation. I don't know - and would honestly love to be wrong - but I would put good money on a prediction that he is equally ignorant, yet an ardent Climate Believer. I know the problems with guilt by association, but do you really want to make common cause with such as he? I cling to the illusion that even in these dark days Richard Lindzen knows more about the climate than Greta T. 'Could be wrong, of course, but I cling nonetheless.

Re India - why the average per person? I suppose it's because "per" is the Covid *mot du jour*. But I see where you're going ... looks like the answer's "fewer persons good" after all. And there was me thinking it was building nuclear power stations.

To O 28 Nov 2021.

A trip to see you would be lovely and might even lift the blanket of misery that afflicts la famille K at the moment. Or, at least, lift a corner of the blanket. But it is something that demands much thought and discussion, including with Caesar. It would also involve discussion with one of the two mechanical wrecks we term "cars", both of which would expire long before we reached you. Anyway, at the moment we are both too frightened to go very far for fear of la peste (a first order fear); or for fear of bringing la peste to others (a second order fear); or for fear of others believing that we come bearing la peste to infect them (third); and so on unto the ends of the world. I have been thinking of buying one of those large copper diving suits with a little glass window to look through. You can tap the glass to see whether I'm breathing.

A few reviews of my Ransome book have appeared. All except one have been quite approving. The exception was written by a woman I thought was a friend. But you live and learn, as Mother said.

To O 28 Nov 2021.

Ever since "climate" began to corrupt science, a misguided application of the precautionary principle has meant a descent into madness. It's application in this Covid era has been a disaster. Here's my rough translation of the editorial (in *France Soir*) I referred you to.

> Mr. President, you have locked yourself into a failing authoritarian strategy, opting for uncertain "vaccination" rather than for the prevention and treatment of disease. By opting for the "Big Pharma" notion of a permanent vaccine "subscription" you ignore the Hippocratic oath ("first, do no harm").
>
> In your defence, you are not the only leader to have acted so calamitously - far from it. But the fact remains that in France, you are the one responsible for one disastrous decision after another:
>
> - at the outset, by refusing to close borders and test patients (including even those repatriated from Wuhan);
>
> - by banning hydroxychloroquine, by refusing to talk about vitamin D, by recommending paracetamol, and leaving Care Home residents to be finished off with rivotril ;
>
> - by subjecting the population to house arrest and then doing it again

in the following winter, with colossal collateral damage and without the slightest tangible impact on the epidemic.

Despite these culpable errors, we never despaired of your coming to your senses - until your speech last night, that is. Now it's clear that you will never do the right thing. Whether you've been badly advised or not, there is no longer any doubt that you cannot be counted on to get us out of this "Covidist" madness.

With you, Mr. President, we have the worst of both worlds: ineffectiveness against the epidemic, and endangerment of our fundamental freedoms. With you, we have experienced the progressive destruction of our hospital system, the all-powerful reign of Big Pharma, and profound divisions in our society, between the vaccinated and the non-vaccinated, between those vaccinated with two doses and those vaccinated with three doses, etc.

You won't get away with this, Mr. President. History will judge you harshly.

Hot stuff, but hard to disagree.

To P 18 Dec 2021.

Life here in France is far from ideal, with many restrictions on daily life and liberty, I guess it is much the same with you. After a relatively long period in which I simply ignored the "covid" affair, over the past few weeks I have found myself becoming increasingly anxious. It's hard to provide a rational explanation for this. Partly it is the way the State has encroached on daily life - all too familiar to those who have lived under a Communist regime but new to me. But also, the rather sinister agenda that old people (that's me for sure!) are surplus to the needs of society and should simply be killed off. Difficult to find that at all reassuring. It is not the first time the medical profession has been tempted down that route.

My book on Arthur Ransome was well received with some good reviews. The Arthur Ransome Society asked me to write a tribute to celebrate the 80th anniversary of the publication of one of his books.

It was quite a lot of work (with some references to Jung) and prompted me to think about a second edition. I took part in a virtual meeting of the Society of Authors in Nice. I did my best, but I dislike trying to interact via a screen. I still await contact from the student Psychology Society at my own university.

To O 20 Dec 2021.

Here is the "GBR" thank-you message.

Dear Friends.

From the depth of our hearts, a belated thank you for signing the Great Barrington Declaration. With over 850,000 signatures, together we opened up the pandemic debate. While many governments continued with their failed lockdown and other restrictive policies, things have moved in the right direction. For example, most schools have re-opened, most countries prioritized older people for vaccination and Florida rejected restrictions in favor of focused protection without the negative consequences that lockdowners predicted.

While occasionally censored, we have not been silenced. Since authoring the Declaration in October 2020, the three of us have actively advocated for focused protection through social media, op-eds and interviews on, for example, vaccine passports and natural immunity.

We have also launched Collateral Global, a charity staffed with academics from across the world to document and disseminate information about the collateral damage of the restrictive measures so that we don´t repeat the mistakes of this pandemic and are able to inform future policy with evidence and analysis. Collateral Global is crowdfunding so that this work can be done to the highest possible standards. You are welcome to join us and help us in those efforts as well as follow us on Twitter, etc.

We are also planning an initiative on scientific freedom soon.

With enormous gratitude.

Jay Bhattacharya Sunetra Gupta Martin Kulldorf.

To P 22 Dec 2021.

Extraordinary post by Will Jones, currently the first article on Toby Young's "Daily Sceptic" website. Half of the comments are from medics. It's basically about "vaccination" driving "cases". Well worth reading the sensible comments. I think the UK is in big trouble.

Life in France is about to become very difficult. The Health Minister, rather cleverly, has proposed a bare bones "vaccine passport" bill and left it to the Assembly to vote changes to it. Devil you do, devil you don't, tricks. French politics is far too hard for me - the UK system seems childish by comparison. All I understand is that while M Macron may well win the election, he could end up with virtually all of his "party" disappearing from the Assembly. He then becomes a figurehead for the conservatives or the LePenites or god knows what.

Strange times.

To P 23 Dec 2021.

Isn't it the case that this virus is doing what viruses tend to do and what we were all told it would do last year? (Oddly enough, this fact itself might be evidence against the lab-creation theory. The journalist Nick Wade speculates that this might be simply a disease brought from the bat caves and handled incredibly carelessly).

I have to confess (if that's necessary) I joined the GBD crowd at its inception. I still think their idea was broadly correct. However, I would temper that conclusion now, because the evidence that "vaccinating" is equivalent to "shielding" starts to look very weak. A vaccine that doesn't do much, has absolute efficacy of 1 percent, wanes rapidly, and is associated with very high levels of adverse effects? Gupta seems to have almost broken ranks over this because (a) she seems to be a socialist, albeit a nice one and (b) she's at Oxford and somebody could well have had a word with her. The evidence from Japan (why isn't it in the news?) seems very convincing that this is a disease that should be treated rather than "prevented".

I'm half way through Robert Kennedy's book. Badly written and massively over-referenced, but you read it and realise he's right. I hesitate to impose it, but you should at least read the first half. He makes the case this could be AZT all over again. If so, it's a complete scandal and there may be interesting consequences. Of course, it's different in France because Swine Flu and contaminated blood were so recent.

To P 24 Dec 2021.

 Read what bit of the review of Robert Kennedy's book that I could, then reached the reference to "de-wormer" and knew where it was coming from. It is simply a hit piece. I doubt the author read the book. It is certainly over-referenced (I guess to cover the libel issue) but it is genuinely interesting (and there has been no libel suit). It is not particularly directed at Gates, by the way: Fauci is the man in his sights. And having looked at some of the AZT references (about which I knew nothing) it is quite extraordinarily persuasive. But you'd need to read the book to have a sensible discussion (it's on Kindle). I have finished it now. (I'm reading *Barchester Towers* for Christmas).

Warning: I'm getting to dislike the term "conspiracy theory/ theorist." It is lazy thinking (and misuses the term "conspiracy"). It was popularised by Lewandowski in the climate context. An effort to characterise people who disputed his (dare I say minority?) world view as "bad people". He sounded a bit paranoid to me, but who was I to say? I would have happily sued him for characterising *me* as paranoid for protesting his truly absurd "recursive fury" paper but fortunately it was retracted in the face of people with deeper pockets than mine. All rather quaint now we know of his own efforts to suppress Dixon et al. Wrong sort of conspiracy I suppose ...

To O 27 Dec 2021.

The statistics in the "Daily Exposé" you cite are completely useless. Sufficiently so for me to doubt the site - maybe it's even a False Flag operation to discredit the sceptical cause.
There is a scandal, but it is not this one. The true scandal is:-

1. The definition of "unvaccinated" is absurd. Some of the more dire after-effects of these mRNA drugs appear within hours or days of injection. Yet subjects are counted as "unvaccinated" until 14 days have passed. We need to know the status of people in hospital counted as "unvaccinated" - have they ever received a dose of vaccine? If so, when? We also need to know why this information is not available.

2. Health Authorities are going to great lengths to disguise the denominators in these calculations. Possibly because in some places (like the UK) they simply don't know the total size of the population! As some of the commentators correctly point out, if nearly everybody is vaccinated then any effect you pick will be greater in the vaccinated group. You require estimates "per relevant population" and you can't know these if you don't know the absolute numbers of "unvaccinated". These data are being suppressed - which I grant might be seen as sinister, but in any case, makes all "vaccinated vs unvaccinated" comparisons meaningless.

Professor Fenton has produced some ingenious analyses to get round this, but they are quite technical. His main point is that non-covid deaths appear to relate to covid vaccination status! Obviously, this makes no sense at all unless there is a systematic source of bias in the data set. He suggests this is the 14-day rule and has some suggestive graphs to illustrate the point. Google Fenton Covid Statistics.

Note, you can draw some (weak) global conclusions from the total numbers apparently getting vaccinated. In France it looks as if about a third of the population has taken all three injections (I don't trust the absolute numbers, but the relative standing of 1, 2, and 3 gives you some idea). That's the problem with the *Pass Sanitaire* - once it penalises a majority of the population there are political consequences. My guess in France is that resistance will come from 18-30 yr olds who like clubbing.

3. The Death statistics are corrupt and include anybody who died

from any cause while "testing positive". In fact, over a third of all covid infections are acquired in hospital. One way round this is to look at "all cause" mortality, available on the Euromomo site. It is quite normal in Scotland and *below average* in France (which would be strange if there really was a pandemic running amok).

To A 28 Dec 2021.

Each covid edict appears more bizarre than the last and we both now live in a state of permanent anxiety. Like the rest of France, you'll no doubt say; nonetheless it's distressing to find a country we love proving so fickle a friend. We simply can't imagine living somewhere where regular – and pointless - "injections" - whether you want them or not - are the order of the day.

We live like three mice in a biscuit tin and see nobody for days on end. Sometimes I stand at the bottom of the hill waiting for Caesar to come back from his "walk" and the whole valley is so completely silent it is scary. And no, it's not because I'm deaf. No cars pass, no people, no hunters, no dogs. The life has gone out of the place.

Thank you for the annual diatribe - It is always greatly appreciated and was read with pleasure. A response will arrive in due course.

To O 3 Jan 2022.

No! Tell her she should not weep for lost times. It is those responsible for this tragedy who should weep.

Index

By the same author

Fiction

> The Boat in the Bay
> The Broken Bell
> The Pink House
> Lucy
> A Time to Tell Lies
> The War and Alex Vere

Non-Fiction

> The Psychology of Reading
> Oscar & Lucy (autobiography)
> A Thoroughly Mischievous Person (biography)
> Why My Mother Went Away (autobiography)

About the author

Alan Kennedy is Emeritus Professor of Psychology at the University of Dundee. A graduate of the University of Birmingham, he has a degree in English Literature and Psychology and a PhD in Psychology. He is a Fellow of The Royal Society of Edinburgh, an Honorary Member of the Experimental Psychology Society and a member of The Society of Authors. He has held academic posts in the University of Melbourne, The University of St Andrews, Monash University, Université Blaise-Pascal, Clermont-Ferand and Paris Descartes University, Boulogne-Billancourt. A founder member of The European Conference on Eye Movements he is an international authority on the cognitive processes underlying skilled reading and is the author of over a hundred journal articles on the subject. The author of seven novels, he has also written biographies of the psychologist Oscar Oeser and the childrens' author Arthur Ransome. Alan Kennedy currently lives with his wife Elizabeth and dog, Caesar, in South-West France.

www.ingramcontent.com/pod-product-compliance
Lightning Source LLC
Chambersburg PA
CBHW032002190326
41520CB00007B/326